Table of Contents: Water Rockets

"Hello, let me guide through the book."

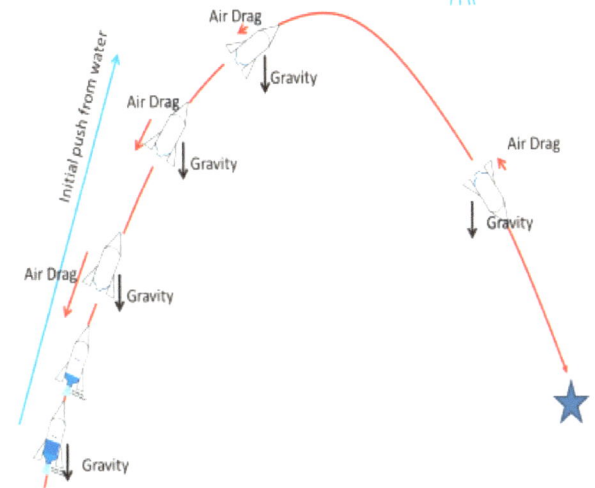

Air Drag — Gravity — Initial push from water

Homemade PVC launcher

Narrow nozzle launch

Metal, high pressure launcher

Low pressure launcher

This book is one of two versions about water rockets.   This abridged version is lighter and talks about general ideas, without going into any engineering analysis. This lighter version is more fun and appropriate for someone who is not interested in engineering equations. If you want the engineering analysis, then read 'Water Bottle Rockets, for Parents and Kids', which is for parents, kids, and engineers.   The two versions are mostly the same except the heavier version has appendices with the analysis.

| Water bottle examples | Toy chemical rockets | Space rockets |

Color key to rocket examples

Exhaust in any form creates a rocket, using pushed water, or hot steam, or a hot gas from a chemical reaction.

Water bottle rockets use air pressure and may not get to space, but these bottle rockets are in themselves a great introduction to toy chemical rockets and space rockets. Air and Water Rockets can be launched many times, are built from scratch, and are a great hands-on start to learning about chemical rockets.

"Same ideas for all the rockets: push something out, and get push-back."

Rockets are very general. Anything that squirts some exhaust out the back is a rocket. For your standard high profile space rocket, a chemical rocket, that exhaust needs its own fuel and oxidizer. For a water rocket, that exhaust needs a bicycle pump, a bottle, and some water, along with some sweat pumping up the compressed air.

Hopefully there is enough thrust or force to lift the body's full weight against gravity, although that is not required if you are going sideways in a rocket car or 'wingman' and just want to fly along the ground.

1. **Water bottles** can easily be made into air pressure powered air or water rockets. To get a lower pressure launch, one kit will give you corks or bungs that pop out when the pressure gets high enough. To get controlled higher air pressure, other kits will provide a PVC pipe launch tube, with tire pump nozzles, and corks.   These launch tubes take more effort to build than the bottle rocket.

2. **Huge space chemical rockets** have the same rocket thrust concept as water bottle rockets, except liquid fuels and oxidizer are used like very cold hydrogen and oxygen.   Solid fuel is also sometimes used as a first stage booster, when the launch needs a large thrust and does not need to throttle.

Water bottle rocket using bicycle pump for pressure

Squids in the ocean

Steam and water exhaust:

The most famous steam rocket was the Skycycle built for Evel Knievel and his jump across the Snake River canyon in 1970. The superheated steam did get Evel across, but then a parachute misfired too early and the rocket drifted back into the canyon.

Sea animals also use rocket thrust. Some octopus can eject a burst of water and escape from predators.

Items around the house from the house plumbing also must obey all these thrust phenomena, just like rockets. When you turn on a water faucet, there is thrust. The faucet jumps back a little from the recoil, although the water flow rate is rather low and made lower using water filters.   When you spray a water hose, there is recoil, again also very small but the phenomena is there just the same.

Test firing of steam rocket

Hot gases from chemical reactions:

If you really want to get into space, the only way is to use chemical reactions. There really is no comparison between the energy density allowed by a chemical reaction and the energy density of compressed air.   Water bottle rockets do demonstrate the same thrust concepts, but at a much smaller energy level.

SpaceX Falcon Heavy

The rocket concept is so fundamental that different fuels and different size rockets are made.

Rocket Types, Air Pressure, Steam, and Chemical Fuel

Exhaust in any form creates a rocket, using pushed water, or hot steam, or a hot gas from a chemical reaction.

Water bottle rockets use air pressure and may not get to space, but these bottle rockets are in themselves a great introduction to toy chemical rockets and space rockets. Air and Water Rockets can be launched many times, are built from scratch, and are a great hands-on start to learning about chemical rockets.

"Same ideas for all the rockets: push something out, and get push-back.

Look at bottle rockets, toy chemical rockets, and then space rockets"

Air and Water Bottle Rockets

Narrow nozzle

Wide nozzle

Air and Water Fuel:
with different size nozzles.

Toy Solid Fuel Rockets

Chemical Fuel:
Solid Fuel Rocket

Space Rockets:
NASA and Private Space Liquid Chemical Rockets with Liquid or Solid Fuel boosters

Liquid Hydrogen and Oxygen (can throttle)

Solid Fuel boosters (burns steady)

SpaceX Falcon Heavy

Space Shuttle

Chemical Fuel:
Liquid Fuel Rocket, assisted by liquid or solid fuel boosters during initial launch

Saturn V for Apollo to Moon

Air Pressure: Air-pressurized bottle rockets can have low or high thrust, depending on how big the hole or nozzle is.

- Compressed air is good for demonstrations, but air does not have enough energy to get to space.

Solid Fuel: Toy solid fuel chemical rockets typically use solid chemical engines because solid fuel is easy to store. The typical fuel amount in a toy chemical rocket is not enough to get to space.

- The beauty of solid fuel is its long term storage, low cost, and high thrust from huge burning surface area. However, solid fuel can not throttle.

Liquid Fuel: Space rockets use a combination of liquid fuel core (2nd and 3rd stages), and liquid or solid fuel boosters for initial launch (1st stage) assistance.

- The beauty of liquid fuel is its better thrust to weight, and ability to throttle the thrust using fuel pumps. In contrast, the beauty of solid fuel is its larger thrust and lower cost.

The rocket concept is so fundamental that different fuels and different size rockets are made.

Water Rockets, Squids, and Plumbing

These air-pressurized water bottle rockets don't require dangerous chemicals. These bottles also don't cost much, because the 'fuel' is you sweating and pumping up the air pressure in the bottle. After the bottle is released, the air pressure pushes out water in a dramatic launch, with the exhaust water and exhaust air causing thrust.

The blast off can be amazingly quick and strong, if the nozzle is wide at 1 inch, or the blast off can be slow and gradual, if the nozzle is narrow at less than 0.2 inches. Either way works.

Firefighters feel the push-back from high pressure fire hoses, and this bottle rocket is the same force. Nature also has animals that squirt out water to move, like squids and jellyfishes.

"Water can push squids, bottle rockets, and crazy people...just squirt water out the bottom."

Water shoots out the bottom, and pushes the rocket upward.

Large nozzle, fast flow, more thrust

Large wide nozzle: Sudden blast off with large flow and large thrust to final speed:
- more water weight is allowed due to large thrust.
- water filled to 1/3rd plastic bottle height, with 100 psi air added with a PVC tube and bicycle pump.

Small narrow nozzle: Gradual blast off with less flow and less thrust to final speed over a longer time:
- less water weight is allowed, due to lower thrust.
- water filled to 1/10th plastic bottle height, with up to 80 psi air added using commercial balloon feed and bicycle pump.

Narrow nozzle, longer flow, less thrust

Fun Facts about different things that use water blasts

Nature:
'Squids, octopuses and jellyfishes, for example, can fill part of their flexible bodies with water and force the water out through a smaller opening, propelling them through—and even out of—the water.'

Squids in the ocean

Water Jet pack:
Think of some crazy person using a water jet pack (a wild adrenaline junkie). The water is blasting out fast, and the pressure comes from a pump floating on the water. As long as you suck the water in at a slower speed than you eject the water, you get a net thrust.
There is more than 200 lbs-force, enough to lift a person.

Water jet pack, like a strong continuous faucet

House Plumbing:
Water squirts out of the faucet really fast. Put your hand right under the stream of water. The push you feel against your hand is the same push (equal and opposite) that the faucet feels in the other direction.

Force back

Water exhaust

Faucet in house

A rocket can blast off, or just propel something along. Even in household plumbing, there is still a backwards force when water is expelled from a faucet.

Daredevil Evel Knievel Jumps Snake River Canyon in Steam Rocket

Using a Steam Powered Rocket, with high pressure and very hot steam exhaust, Evel Knievel, the motorcycle daredevil in the 1970s, jumped the Snake River Canyon. Hot high pressure steam is an extreme example of a water rocket.

Super heated steam has much more pressure than your standard water bottle rocket. Evel Knievel is part of popular culture and folk lore, the beginning of xtreme sports.

"What a crazy, fun guy. His team built a super steam rocket."

Steam Rocket

Super heated, high pressure water

Test firing of steam rocket

1600 FEET

Blimp filming the jump

Evel Knievel Fails Snake River Canyon Jump

Why not use a chemical rocket? Steam is good enough for a short hop.

Water was superheated in a closed sphere. To start the launch, a valve was opened, allowing water to escape, vaporize, and shoot out the back with huge velocity. There is much more energy in superheated water than in compressed air.
For the rocket hardware, the Steam Powered Rocket was welded together with a small team of people. Unfortunately, the rocket had a mechanical failure and the parachute deployed by itself early, stopping a successful crossing. The rocket actually got across, but then drifted backwards into the canyon after the parachute opened.

Evel Knievel's Steam Powered Rocket Cycle

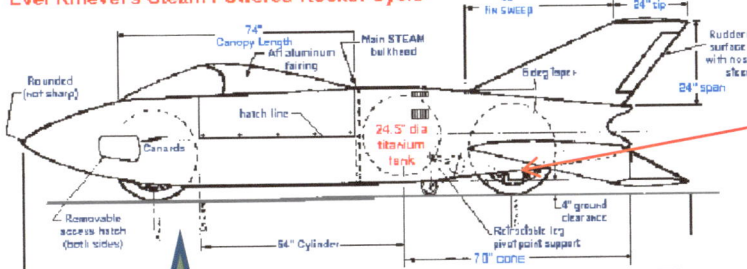

45° FIN SWEEP
24" tip
74" Canopy Length
Aft aluminum fairing
Main STEAM bulkhead
Rudder control surface moves with nose wheel steering
24" span
Rounded (not sharp)
hatch line
Canards
24.5" dia titanium tank
Removable access hatch (both sides)
Retractable leg pivot point support
1.4" ground clearance
54" Cylinder
70" cone

Experiment: boil water on rollers, and let the steam push the rollers.

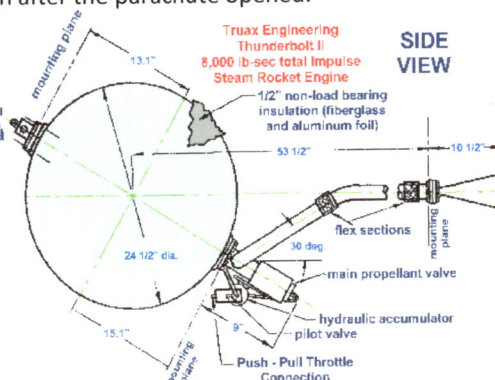

Truax Engineering Thunderbolt II
8,000 lb-sec total impulse Steam Rocket Engine

SIDE VIEW

13.1"
1/2" non-load bearing insulation (fiberglass and aluminum foil)
53 1/2"
10 1/2"
24 1/2" dia.
mounting plane
flex sections
30 deg
main propellant valve
9"
15.1"
hydraulic accumulator pilot valve
Push - Pull Throttle Connection

8000 lb-sec Impulse (~5 g's for 1 second, to get to 50 m/s rocket speed)

Poseidon rocket:
A submarine rocket uses steam to eject from the launch tube, before chemical rocket kicks in.

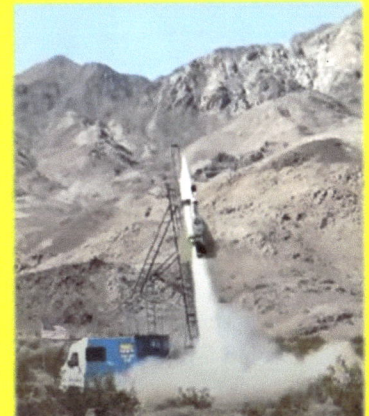

Rocket Man in 2018:
Rocket using high pressure steam and steam exhaust to blast up 2000 feet.

Knievel's steam rocket needed a large valve to release the super heated steam.

Design Team Behind the Knievel's Steam Rocket

It takes a team to design all the rocket parts to get a huge thrust from super high pressure steam to cross the canyon. Who designed the rocket?

Three Skycycle X-2's were built. The third one was used to make the launch with Knievel inside, in the early 1970s. The first one dove into the Snake River. With design changes, the third one was built more like an airplane than a motorcycle.

Knievel's rocket was designed by Robert Truax. Truax development many rockets for the military from the 1940s to the 1960s, in the Navy and the Air Force. He helped discover a chemical fuel that spontaneously burst into flame when combined with nitric acid.

In the late 1960s, Truax retired from the military are founded Truax Engineering, to continue to advise and develop rockets.

Truax was very familiar with steam thrust because military rockets would use steam to eject chemical rockets or torpedoes from launch tubes.

Truax's goal was to build a reusable space tourism rocket, even around 1970. Well, it has taken 5 more decades than planned, but space tourism is now starting to become a reality, for the very qualified or the very rich.

Evel Knievel's X-2-2 Skycycle on display at the Harley-Davidson Museum in 2010

Robert Truax designed the Skycycle X-2.

Eddie Braun successfully recreates Knievel's jump in 2016, using a replica of the same steam powered rocket. Truax's son designed the replica rocket. The only change was a better parachute, which had faultily deployed in the 1970s jump.

Stuntman Eddie Braun successfully recreates Knievel's jump in 2016.

The rocket reached approximately 400 mph after launch, and the jump covered more than 1,400 feet.

Rocket just before re-enactment launch in 2016

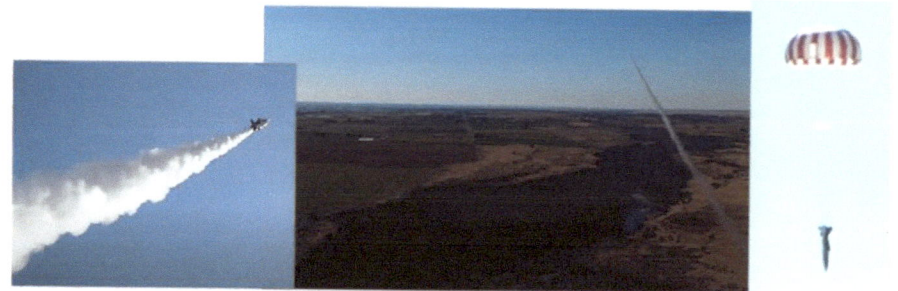

Steam launch, and parachute descent

The thumbs-up success

A very experienced rocket engineer developed Knievel's steam rocket.

There are many launch pads to launch bottle rockets, either homemade or ready to buy. The launcher lets you seal the bottle, lets you pressurize the bottle, and lets you release the bottle. The water exhaust and air exhaust blasts the bottle up to the tree tops. The water and air exhaust cause a thrust.

Let's use air pressure! Air pressure is great for bottle rockets because the stored energy of the high pressure gas can be released right away. It might take a slow minute of pumping with the bicycle pump, but during launch the gas is released in fractions of a second to get large water exhaust velocity and get large power, like a chemical explosion although not as powerful.

Homemade launchers typically use PVC pipe with a bump pushed into it by heating a small section of pipe and then pushing in from both ends. Another way to seal the pressure is to use a cork around a PVC pipe. And there are other ways too, like o-ring seals, or just a bung and hose without a launcher.

Thrust
'Reaction'

Pressure

Exhaust
'Action'

'Reaction' from pushing out water and gas

Other common uses of air pressure:
The application of air pressure for bottle rockets is the outlier use of air pressure. There are also many everyday uses of air pressure. Let's acknowledge water fountains, air pressure in car and bicycle tires, and garage air pressure tools.

Household plumbing uses air pressure to keep the water pressure in the pipes. The pipe pressure is always trying to push the water out of the faucet or shower head. What if we didn't have the air pressure idea for plumbing? The alternative would be a water tank high above the house on a hill, like in some towns. Another alternative would be to turn on an electric pump every time someone wants to use the water, but that would use a lot of electricity.

Yes, there is an upwards force on the faucet when water is pouring out.

Less common uses of air pressure:
There are also some non-typical engine applications, such as using air pressure to power a piston engine. The energy stored in a pressurized cylinder is not the same as an old fashioned steam train engine that uses water and coal, but the engine just might get you around town for a day before getting re-pressurized.

One ride at the fair, the 'Screamer', uses the idea of a fast release of air pressure to launch people up a tower.

The idea that air pressure can push on things is given by the ping-pong cannon, where a vacuum is pulled on one side of a ping-pong ball. When the other side is exposed to typical air pressure, the ball shoots out faster than the speed of sound and can rip through soda cans.

Car Tires

Pump up the plastic bottle with high pressure air. Release the clamp holding the bottle down against a seal. The high pressure air pushes water out and the rocket gets pushed in the opposite direction.

Air and Water Pressure in Bottles and All Around Us

There are many launch pads to launch bottle rockets, either homemade or ready to buy. The launcher lets you seal the bottle, lets you pressurize the bottle, and lets you release the bottle. The water exhaust and air exhaust blasts the bottle up to the tree tops. The water and air exhaust cause a thrust.

Homemade launcher from PVC pipes, with bump seal or rubber cork

80 psi launch from homemade wide nozzle PVC launcher, with bump seal

3...2...1...blast off

water expelled

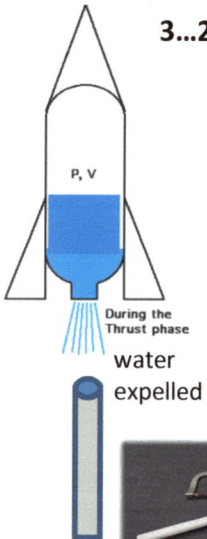

Make your own bump and pipe connections

Kit launchers you can buy, with reliable seals and release mechanisms

PVC kit launch at 80 psi

Relationshipware StratoLauncher at 80 psi

Boom!

25 psi launch from wide nozzle kit 'Aquapod'

PVC Creations kit with Bump in PVC pipe: wide nozzle with release

Medium nozzle with metal o-ring seal

Wide nozzle with o-ring and release lever.

Stand alone launchers you can buy, with self release corks

Woosh!

80 psi launch from narrow nozzle kit 'Antigravity'

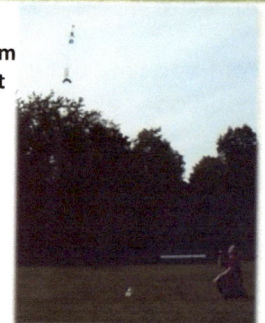

2 stage water rocket kit

Narrow nozzle

Two stage

Bottle with tight cork

Pump up the plastic bottle with high pressure air. Release the clamp holding the bottle down against a seal. The high pressure air pushes water out and the rocket gets pushed in the opposite direction.

Air and Water Bottle Rockets: The Basic Idea with Clamp

Bottle rockets are just a pressurized bottle pushing out water and air as exhaust and thrust, with the same thrust concepts as space rockets.

Bicycle pumps can get up to 100 psi, which is a safe pressure for plastic bottles with thick walls.

"Deep down in their gut, people get this. Just need a tight seal, and need a pipe going above the water level.
Water rockets can launch with a bang!"

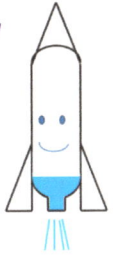

Technique 1: PVC launch tubes with a bump or seal. You have control over the release pressure using a clamp. The rocket is held down against the seal by garbage ties, until released by pulling the rope.

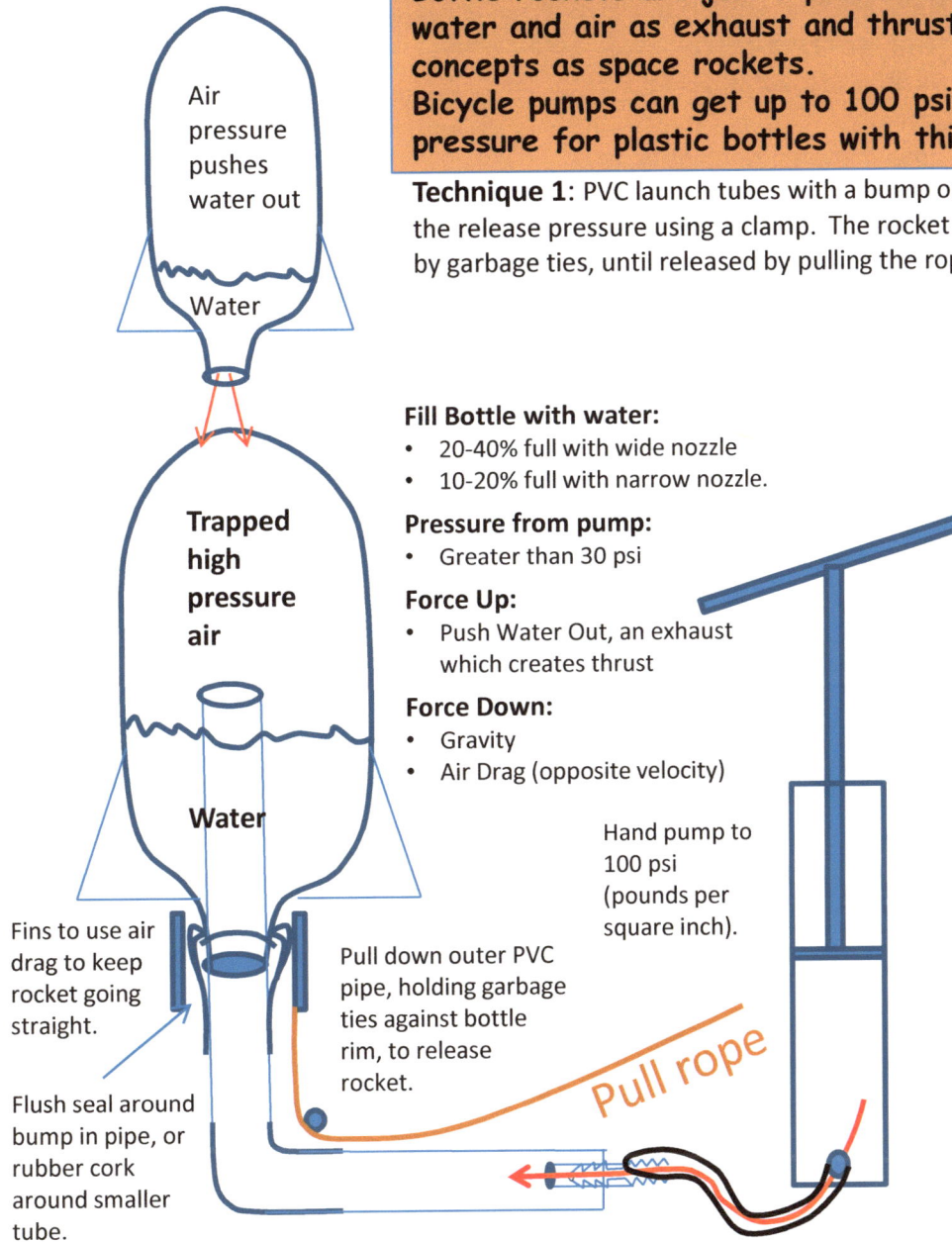

Air pressure pushes water out

Water

Trapped high pressure air

Water

Fins to use air drag to keep rocket going straight.

Flush seal around bump in pipe, or rubber cork around smaller tube.

Fill Bottle with water:
- 20-40% full with wide nozzle
- 10-20% full with narrow nozzle.

Pressure from pump:
- Greater than 30 psi

Force Up:
- Push Water Out, an exhaust which creates thrust

Force Down:
- Gravity
- Air Drag (opposite velocity)

Pull down outer PVC pipe, holding garbage ties against bottle rim, to release rocket.

Hand pump to 100 psi (pounds per square inch).

Pull rope

Water Rockets

Hey Scouts, water rockets can be part of a space exploration merit badge!

Tips for a successful group launching activity, by blasting out water:
Fill the bottle with ¼ water, and pump up the air pressure with sweat and muscle. Water is the thrust for longer thrust than just air.
Each scout made their own rocket from plastic bottles, and cut the fins from poster board.
The scouts are waiting, to the side out of harms way, to launch their water rockets.

Water Stop

Water starts

Bottle rocket event for scouts

Air Rockets

Air rockets use the same launch setup as a water rocket. Don't fill the water and you have yourself an air rocket. Air is the exhaust!

Pump a bottle rocket using bicycle pump

Rocket and bicycle pump

Rocket stuck in a tree

Here is the basic idea for pumping up and launching a bottle rocket.

Household Uses of Water Pressure, That You Depend On Every Day

Have you ever seen faucets jerk up a little when you turn on the water, especially if there is no mesh filter at the tip slowing the water flow? The water spraying out is an exhaust, and it must provide a thrust backward, just like a rocket. Newton said this is 'action' and 'reaction'. Here are some Q and A about the air and water pressure that is probably in that wall next to you in pipes.

"Water, water, everywhere, all around you in the wall plumbing."

House plumbing:

- **Think of a house faucet.** Water squirts out of the faucet really fast. If you turn the water on quick, you might see the faucet jerk up from the force at the nozzle. Now, there might even be a little mosquito-mesh-like filter at the tip of the faucet, but the water still squirts out really fast. Put your hand right under the stream of water. The push you feel against your hand is the same push (equal and opposite) that the faucet feels the other direction.
- **Think of an outdoor water hose.** Hold the lever and feel the kick backwards as you quickly turn on the water spray. OK, there is not much of a kick from a garden hose, but the nozzle does try to rotate up due to the backward force of the water.

What does this high pressure and fast spray mean for houses? —Faster water in showers, and water reaches the upper floors against gravity.

Generating water pressure in house pipes:

- There's a lot of pressure, more than 50 psi, inside the plumbing of the house, and that pressure is generated by either a well pump or high pressure air tank inside the house, or by the town water stations, or by the huge water towers 50 feet above any house somewhere at a hill in the neighborhood.

Pressure needed in house pipes:

- The pressure in house water pipes is typically above 50 psi in the basement, where water takes 14 psi to go up every 33 feet. The pressure on the 3rd floor is 14psi less, and you still want plenty of pressure left over to have a good spray. A three story house needs more water pressure than a 1 story house. Houses do have strong metal pipes to hold the pressure easier, instead of a plastic bottle.
- Water bottle rockets at 100 psi don't have much more pressure than your house.

Yes, there is an upwards force on the faucet when water is pouring out.

Force back

Water exhaust

Kitchen faucets are static rockets

Copper pipes in homes

Water hits tree

Lower pressure garden hose

Higher pressure, high flow rate fire hose with lots of thrust

We're surrounded by water pressure all the time, in plumbing pipes in the walls. Turn on the faucet, water sprays out, and there is recoil.

Fun Facts: Long Live Air Power

Is compressed air practical and already used?   YES

Air pressure is used everywhere: inflate car tires, spin shop tools, push water out of a water fountain and water pressure in homes. There are also some experimental uses like a car powered by a high pressure tank.

Again, water bottle rockets use air pressure to push out water as exhaust for rocket thrust or for the air to be the exhaust itself. Still, that is a rather niche use of air pressure. The more typical use below just uses the pressure to push something, like a piston or a turbine.

Air pressure is used everywhere in gas and diesel engines, to heat the gas up before the fuel explodes.

"Compressed air can do a lot more than power water rockets...how about scuba diving, machine shop tools, and pressurize water fountains."

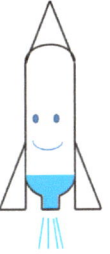

Common uses of compressed air

Car Tires: Your car and bicycle rides on compressed air.

SCUBA Tanks: Highly compressed air (>3000 psi) keeps scuba divers breathing

Water Fountains: Compressed air pushes water out of water fountains

Tools in car garage and machine shops: Pneumatic tools spin using compressed air

Gas driven compressor: Power the tools.

Pellet guns: Compressed gas (air or CO2) pushes out BB's and pellets from air guns

Experimental uses of compressed air for transportation

Power Scooters: Compressed air can power a scooter

Power Cars: Compressed air is used to power some cars in India

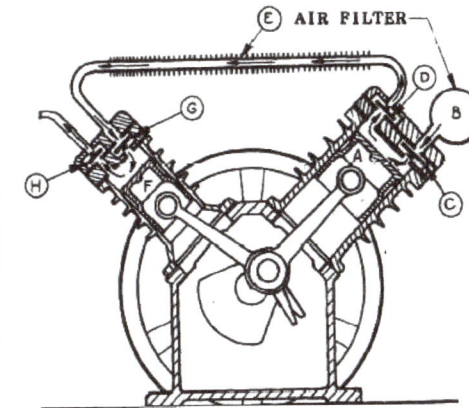

AIR FILTER

Air Hogs pump plane, with pump. The bottle body with air powered piston was used for the propeller.

Pistons powered by air pressure instead of explosions

High pressure air is an energy source.   High pressure air can power a small car, or power hand tools in a machine shop, or push down on water to make a water fountain squirt up water.

"Let's explore 'Action and Reaction'."

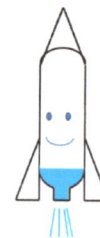

This chapter asks a few fundamental questions.
- How do rockets work?
- What are the basic rules about thrust, from the Newton's laws point of view, by pushing stuff back?
- Are chemical rockets much better than water bottle rockets?
- Who launched the first liquid chemical rocket, and how did it compare to a bottle rocket?
- What other modes of transportation use thrust, by pushing stuff back?
- Why don't cars use rockets for propulsion?
- Does anything practical use water exhaust for thrust?
- Why can't humans fly?
- Do superheroes break the laws of physics?

Who obeys physics?

Ironman, with exhaust

We compare water bottle rockets, toy chemical rockets, and space rockets. Water rockets, for example, have a low exhaust speed, and therefore are 'toys' and only go up 100 meters. Toy solid fuel rockets, like Estes or Quest, have a much faster exhaust speed and are versions of chemical space rockets although just smaller and less efficient, and with much less fuel mass compared to the total weight of the rocket. Space rockets use every trick in the book to get into orbit, from most exhaust speed to very high mass ratio to multiple stages.

Rockets are amazing, and need to have huge instantaneous power by burning a controlled explosion of a chemical reaction. This kind of power is needed to get thrust into space for useful rockets.

Human muscle can't get these kind of instantaneous powers. We as humans can jog a marathon for a long time, but that is a steady but weak power of about 100 Watts of power. Chemical rockets use and need an instantaneous 100s of MegaWatts of power, or even 10s of GigaWatts for space rockets, to get a large enough thrust for large payloads against gravity.

Solid fuel rockets have been around for over 1000 years, either as fireworks or basic rockets in China. Liquid fuel rockets were enabled in the early 1900s because scientists learned to make pure oxygen into a cryogenic liquid.

Some superheroes use rocket power and exhaust to fly, like Ironman. These superheroes are obeying Newton's law of action and reaction, at least in concept. On the other hand, a superhero like Superman has no explainable mechanism to get lift and thrust, because there is no exhaust.

Superman, with no exhaust

Anything needs something to push away to get an opposite force. Rockets need to push something away to get an upward force. Faster exhaust speed gives more thrust.

Thrust and Acceleration When You Push Something Away

Rockets use the idea of 'action and reaction', or Newton's 3rd law. When something is pushed out the bottom as exhaust, an equal and opposite force pushes up on the rocket.

Water rockets, for example, have a low exhaust speed from the limited 100 psi bottle pressure, and therefore are 'toys' and only go up 100 meters. Toy solid fuel rockets, like Estes or Quest, one, have a much faster exhaust speed than water rockets, and, two, are versions of chemical space rockets although just smaller and less efficient and much cheaper.

"Here are real, practical rocket ideas."

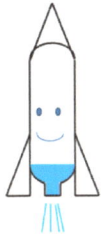

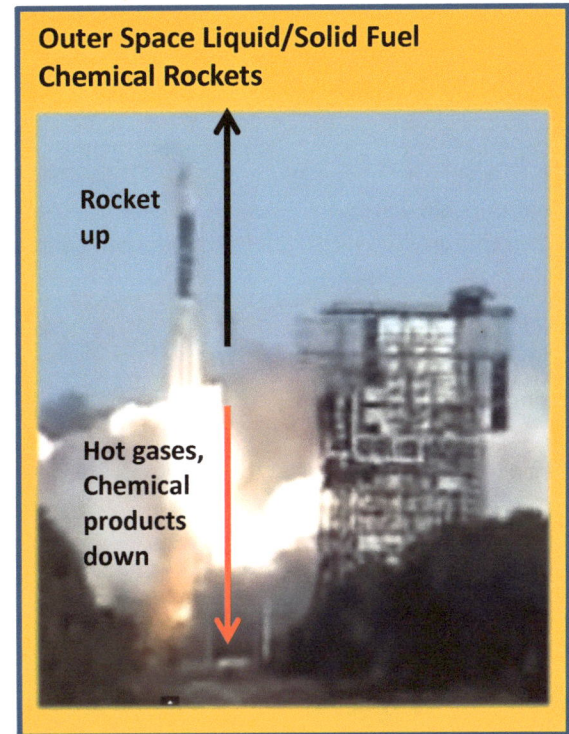

Water Bottle and Gas Rockets

Rocket up, 'Reaction'

High pressure air

Water down, 'Action'

Explosion

Toy Solid Fuel Rockets

Rocket up

Hot gases down

Outer Space Liquid/Solid Fuel Chemical Rockets

Rocket up

Hot gases, Chemical products down

	Bottle Rocket: Air Pressure	Bottle Rocket: Explosion	Toy and Space: Solid Fuel	Space: Cryogenic Liquid and Solid Fuel
Rocket type	Compressed air and water	Explosive gas	Toy Estes and Quest model engines	High mass ratio space rockets, like Goddard's first rocket
Exhaust material	Out goes water and then air	Out goes hot gas: Hot water vapor from hydrogen gas, starter fluid spray	Out goes hot gas: Space Shuttle booster rocket	Out goes hot gas: Hot water vapor from liquid oxygen, kerosene, liquid hydrogen

Anything needs something to push away to get an opposite force. Rockets need to push something away to get an upward force. Faster exhaust speed gives more thrust.

Action and Reaction from Exhaust, How Exhausting!

Both bottle rockets and chemical rockets use exhaust to get thrust. Hot chemical reactions have much more heat and energy and exhaust velocity than compressed air, but the air and water bottle rockets are still rockets. Bottle rockets have exhaust, either water or air. Guns are just another example of rocket thrust, in this case called 'recoil' or 'kickback', by pushing out a bullet.

Rockets Forces:

- Push stuff away from you, like a bowling ball, and you get an equal and opposite force pushing on you (Newton's 3rd law). That makes you use the same 'action and reaction' forces as a rocket.
- The expelled stuff can be anything: compressed air, exhaust water, exhaust hot gas from chemical reactions, or hot gas from accelerating ions, or bullets.

Home Measurements:

- With modern digital cameras, you too can measure all these forces and accelerations, using videos at a regular frame rate or slow motion.

You push on a wall and the wall pushes on you.

Videos will show height versus time, or acceleration.

"Here is the main concept for rocket thrust. When you push something out – Action – there is pressure pushing you forward – Reaction."

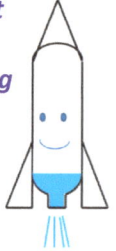

Bottle Rockets thrust:

Push out the cold water just using compressed air.

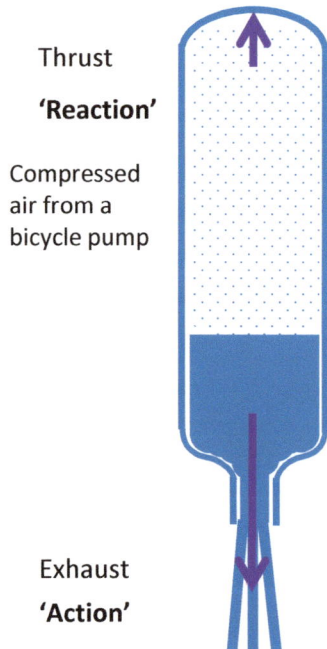

Thrust

'Reaction'

Compressed air from a bicycle pump

Exhaust

'Action'

'Reaction' from pushing out gas:
- The unbalanced pressure on the top of the combustion chamber is doing the lifting.

Chemical Rockets thrust:

Create hot pressurized gas using a chemical reaction, between the fuel and the oxidizer.

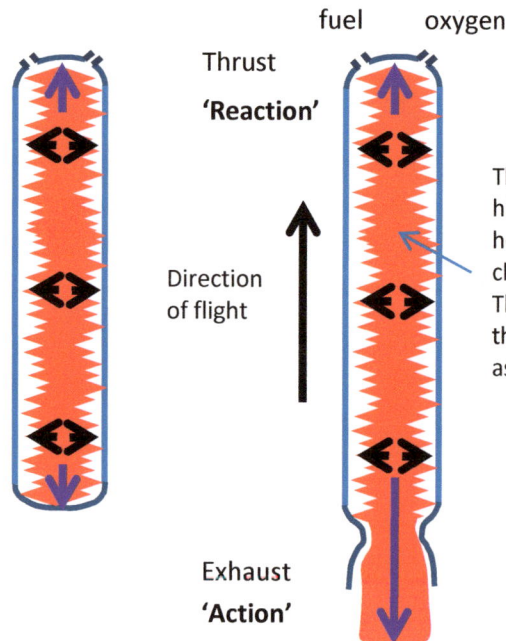

fuel   oxygen

Thrust

'Reaction'

Direction of flight

The gases are much higher pressure for hot gases from a chemical reaction. This high pressure is the same in concept as a bottle rocket.

Exhaust

'Action'

'Reaction' from pushing out gas:
- Gas in a closed container pushes equally in all directions.
- Gas escaping through a nozzle (bottom) causes the container to react and move in the opposite direction.
- The converging/diverging nozzle tends to aim the gas downwards, instead of sideways. This increases the thrust.

Guns and Cannon thrust or recoil or 'kick back':

Guns have a rocket force too, called the recoil. A bullet (exhaust mass) is getting shot out the barrel (nozzle exhaust from bottom).

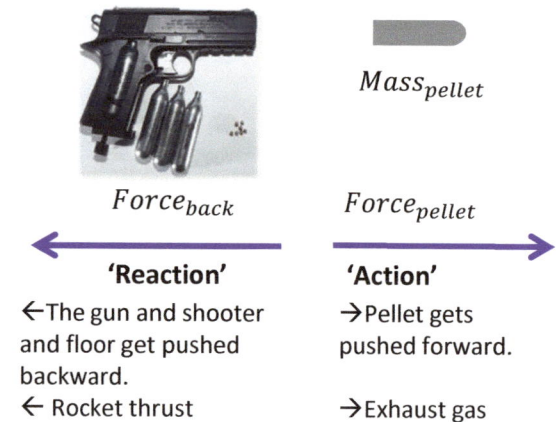

$Mass_{pellet}$

$Force_{back}$ $Force_{pellet}$

'Reaction'   **'Action'**

←The gun and shooter and floor get pushed backward.

← Rocket thrust

→Pellet gets pushed forward.

→Exhaust gas

Recoil is a form of rocket thrust:

- Hand gun jerks backward and up when fired, from the kickback, pivoting around shooter's hand.
- For rifles, the shooter holds the rifle butt against their shoulder so gun doesn't have a gap to fly back and 'punch' the shooter.

For every force backwards, there is a force forwards. For every action, there is a reaction.

Newton's Three Laws of Motion. It's the Law.

Newton described the idea that forces cause motion, and he described action and reaction. Newton's laws form the basis of motion, especially for rockets. He lived over 300 years ago, before even the existence of hot air balloons. Before 1900, people use to say 'why do science, everything has already been discovered?' As we know, a lot has happened after 1900, such as chemistry, electronics, and space rockets.

1. **Newton's 1st law: Things in motion like to stay in motion. An object stays still or continues at same speed unless it is pushed or pulled by something. This is the law of inertia.**
   - True for cars rolling along horizontal ground without friction. Gravity does not slow the car down because the force of gravity is perpendicular to the coasting direction.
   - True for spaceships out in interstellar space, away from Earth's or Sun's gravity.

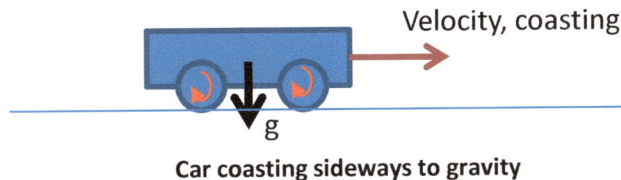

Velocity, coasting

g

Car coasting sideways to gravity

Voyager spacecraft, right now coasting in interstellar space (launched in 1977).

2. **Newton's 2nd law: The force on an object is equal to the object's mass multiplied by its acceleration:**
   Force = mass * acceleration. This is the law of acceleration.
   - True for the forces of thrust, gravity, and air drag acting on the rocket.
   - When you know the force and the mass, then you can calculate the acceleration. Or, when you measure acceleration and mass, then you can calculate the force.

These water rocket accelerations, of 1g's, are comfortable for people if they were sitting in the rocket.

height=4in
T=0 sec

height=15in
T=0.125 sec

height=40in
T=0.250 sec

Launch of water bottle rocket

This space rocket has an acceleration that is human-friendly and tame, of about 1 g, exactly like manned rockets needs to have.

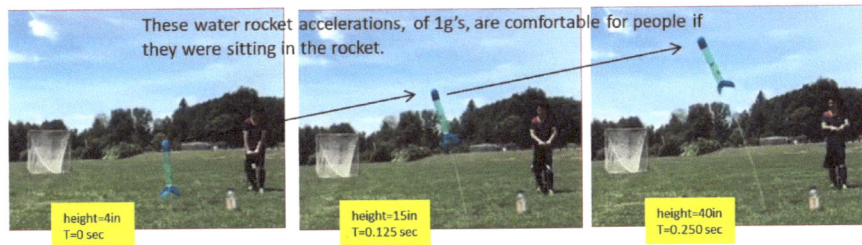

Launch video:
Delta 2 rocket launch of NASA MITEX spacecraft
from Cape Canaveral, FL
Rocket length: 39 meter length (128 feet)

39 m

Height = 0 meters
T = 0 sec

Height = 8 meters
T = 1 sec

Height = 22 meters
T = 2 sec

Height = 40 meters
T = 3 sec

Launch of space rocket at 1 g.

3. **Newton's 3rd law: Action and Reaction: Every force has an equal and opposite force.**
   - True for thrust generated by rocket exhaust.
   - True for propellers on boats, and propellers on airplanes.

Boats: propellers push water faster out the back.

Airplanes: propellers push air faster out the back.

Rockets: Rocket exhaust from Falcon Heavy

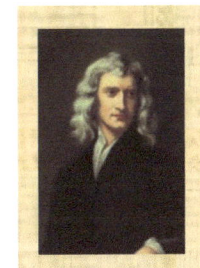

Sir Isaac Newton, 1663: 3 laws of motion

Need rocket exhaust, or something to push away, to get thrust from Newton's 3rd law

Rocket Cars Going Along Ground (...are Silly)

Ever wonder why cars are not powered by rockets? Sorry, car and rocket enthusiasts, rockets have low thrust and very poor miles per gallon at slow speeds. A combustion engine with gears that directly power the wheels is a better idea.

"It's only rocket science... Just add wheels and still Push, Push, Push."

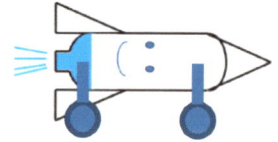

Rocket cars: Same rules for space rockets apply to rocket cars.

Both space and rocket cars have thrust from the exhaust gas.

One good difference is that gravity does not slow the rocket car down, because the car is going sideways. Still, air drag and tire resistance do slow the rocket car down. Another bad difference is that rocket cars can not compete with combustion engine cars. Rockets are only good when traveling fast, which they are great at doing. Rocket engines are not exploiting oxygen in the air, so the efficiency of rocket engines at low speeds is terrible compared to a regular car combustion engine. Poor efficiency means burning through many gallons of gas per second, at the same slow speeds as a car with a combustion engine. Rocket engines at low speeds also have less thrust than a combustion engine or propeller at low speeds for the same fuel.

- For cars, a standard combustion engine is better than a rocket at low highway speeds.
- For airplanes below the speed of sound (sub-sonic), better to use propellers at low <300 mph speeds, and better to use turbo-jets between 300 mph and Mach 1 (700 mph).

Air bottle and toy rocket car

Gas pressure CO_2 car: The CO_2 car can travel without gravity, as a rocket! You can race along the flat ground.

Air-exhaust and CO_2 cars: the push can be air or water too.

Toy chemical rocket engines on derby car

Chemical rocket cars for people

Rocket cars: rocket exhaust moves car forward with no power to the wheels.

Rocket on bicycle: avoid the pedaling, with no power directly at the wheels.

Cartoons knew about rocket cars all along, starting in 1949.

Rocket sleds: The fox gets the built up speed, but he can't turn fast enough to catch the slower roadrunner.
This large turning radius is a real effect when going fast. Slower rabbits escape charging foxes, by zig-zagging around.

Rocket engines can go on cars too, not just on space rockets. Rocket engines are pointless compared to a combustion engines at slow highway speeds, but it is cool.

Thrust and Acceleration, Newton's 3rd law

Any exhaust, especially fast exhaust, causes thrust. Newton's 3rd law is very clear about that. Besides rockets, that law fits airplanes with exhaust out the engines and cartoon Ironman with glowing exhaust, but the law does not fit some cartoon superheroes like Superman. Rockets, or airplanes, need to shoot something out to move forward. Rockets, guns, balloons release air as exhaust, YES.
People hovering by themselves, witches on broomsticks, release nothing, NO.

Real: Newton's 3rd Law

What do birds, airplanes, and rockets push against, in the real world?
- Birds and Airplanes push down and backwards against the air, so the air is the backwards exhaust.
- Rockets have backward exhaust, from burning fuel and oxidizer.
- All these demonstrate Newton's action and reaction in order to move forward.

How to move forward and fly for **birds and airplanes**?
- Push down and backwards on air

air

Bird pushing air down

air

Jet engines pushing air back

Ironman: Extreme heating of air with fusion power source (real in concept, although not in practice).

How to move upward for **rockets**?
- Push exhaust backwards to get thrust upwards

Rocket exhaust

'The truth shall set you free' and actually allow practical transportation to happen, using Newton's 3rd law.

Imaginary: Motion Against Nothing

Superhero Super Thrust: Where's the 'Action' and 'Reaction'?
A Superhero can fly faster than a speeding bullet, and turn and speed up in arbitrary directions, without any exhaust! How?

What are the superheroes pushing against?
Nothing ...against physics.
Something ...Antigravity? Warped space?
- Just for fun, 'antigravity' would propel the superhero away from earth, not sideways.
- 'Warped space' could work if there was a black hole always slightly in front of the superhero, but what is propelling the black hole?
- Maybe a superhero can distort space to create attraction, but this is all unexplained.

There is no 'how'. This is all imaginary stuff.
We need action and reaction in order to move forward. No offense to Superman, but is he farting his way forward?

Superman breaking Newton's laws, with no exhaust

'Ah humbug', or 'What a buzz kill' for those who want to leave reality at the door.

Water has lots of momentum, and It Came From Action and Reaction

Water can pack quite a punch. It can have a lot of momentum, and momentum is hard to stop. The momentum also had to be generated by an opposite force, just like the action and reaction for rockets. When the water bottle rocket is ejecting something as heavy as water, there has to be a strong thrust.

Here are other water momentum examples. Waves on the beach can easily knock people over. Water hoses have been used for crowd control, and to protect ships from boarding hijackers.

Waves:
Waves are generated usually by storms over the ocean. The wind keeps pushing the water surface and causes the turbulent waves.

Waves can also be generated by earthquakes. The ocean bottom can suddenly jerk to the side, or suddenly drop a meter in depth. This causes the water to drop into the extra space, which creates a wave.

Either way, if you get in front of a wave, you can easily be tossed by it, and you can feel like you are fighting a mountain. Generally it is best not to fight the wave. You can swim under it. In a rip tide, you can swim sideways to the flow.

Hoses:
Streams of water can also be used to push people back. These techniques are a form of non lethal crowd control or cargo ship protection.

The water pumps provide the action on the water stream, and there is a reaction force on the water pumps.

A wave can easily knock people over.

Sometimes it is fun to jump into waves, and sometimes you just want to get out of the way.

Regular coastal waves can carry surfers.

Here is another way to use nature's power for fun.

Water sprays are used to stop unwanted boats boarding shipping boats.

Water sprays are non life threatening ways to stop hijacking of boats.

Being pushed around under a wave or in river rapids will make you a believer that water can have lots of momentum.

Steam and Air Ejection of the Poseidon Rocket and Torpedoes ... a little history

Poseidon rocket:
Steam pushes the Poseidon rocket out of the launch tube. The high pressure steam is produced from a solid fuel boiler.

The Poseidon rocket was in service from 1972 to 1992, after the fall of the Soviet Union.

The Poseidon rocket is a two-stage solid fuel rocket. It is launched underwater from a submarine. The main motor is ignited when the rocket is approximately 10 meters above the submarine, still under the water.

So the rocket is a chemical rocket, but it gets a boost out of the launch tube from the steam push, just like a boost for a water rocket from the pressure in the PVC pipe launcher.

The rocket is actually neutrally buoyant in the water, at 64,000 pounds. This rocket is very heavy, but so is the displaced water, and the two weights are about equal. The rocket is very large so it displaces a lot of water, which creates neutral buoyancy. When the main engine starts to burn, the thrust only needs to push through water drag. After the rocket gets above the water, then the full weight of the rocket needs to be supported by the thrust.

This steam push out of the tube is similar to the initial push of the water rocket by the pressure of the launch tube. No exhaust is pushing the rocket up, just the steam pressure from the tube.

However, the steam does need to be high pressure, probably more than 30 psi. The steam pressure needs to lift the weight of the rocket out of the launch tube, when the outside water is not providing neutral buoyancy. With a 72 inch diameter rocket, the steam pressure needs to be larger than 14 psi to lift the rocket, and have extra pressure to push against the water above.

Torpedoes:
Air pressure is used to push the torpedo away from the ship. The propeller on the torpedo can also be pre-spun using steam or air pressure.

The Poseidon rocket uses high pressure steam inside the launch tube which pushes the rocket away from the submarine underwater, before chemical solid fuel rocket engines ignite.
Torpedoes use high pressure air to eject the torpedo before the fuel ignites.

United States Poseidon rocket leaving the ocean after the chemical rocket engines ignite.

The submarine is hidden below the surface.

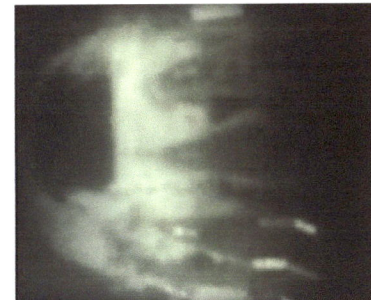

Russian Poseidon rocket launch out of launch tube underwater.

Mark 32 torpedo pushed out of the launch tube using steam pressure.

Water Flow Discovery of Converging / Diverging Nozzle ... a little history

A steam nozzle actually lead to the discovery of the converging / diverging nozzle. Without this shape, space rockets would have much less thrust. The heat energy would not be channeled into an exhaust velocity all in the same direction to create the most thrust.

Converging / diverging nozzles were not first developed for rockets. Instead, these nozzles were developed to make butter using a steam powered spinning machine for cow's milk, before the age of combustion engines and electric motors, by a man named de Laval.

The converging / diverging nozzle was first invented for the dairy industry, to increase the spin of a spinning machine to separate the milk from the cream to make butter. The spinning machine uses steam turbine to focus the steam into a jet, to spin faster.

Cream separator, using a hand crank to spin the milk, in the 1880s.

Use hot steam to spin a shaft, to spin the milk. Electric motors did not exist yet.

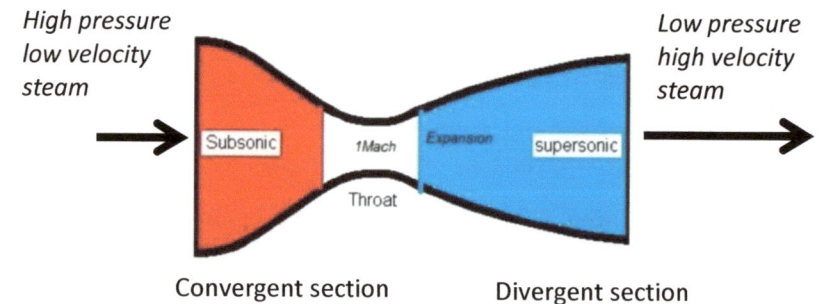

High pressure low velocity steam → Subsonic | 1Mach | Expansion | supersonic → Low pressure high velocity steam

Throat

Convergent section — Divergent section

Diagram of a de Laval nozzle, showing approximate flow velocity.

Pinch the nozzle tips to create convergent / divergent shape, and increase thrust. The spinning torque from hot steam will be larger.

The inventor de Laval did rockets a favor when he improved the cream separator by inventing a converging / diverging nozzle to increase the exhaust speed to spin the cream separator.

This chapter is a survey of 6 launch kits you can buy. These commercial launch kits are compared to homemade launch pad, where you start from scratch and design the launch pad yourself with PVC pipes.

Commercial launch kits use many different ways to build up pressure in the bottle and launch it. Most ways have a bicycle pump with a hose into the nozzle to pump up the bottle, after the water is poured into the bottle. Another way is the fizz from dissolved baking soda or Mentos.

The hard part of the launcher is the pressure seal, and the release lever. We don't want the water leaking out as the pressure gets higher.

"Let's do a survey of convenient launch pads you can but, and compare them to a homemade launch pad."

Let's look at different launch pad designs, either one you can buy or one you can make.
There are many commercial kit launchers, and you'll probably only get one. When you make the decision of which one to get, this is not like shopping for a car, where you get to test drive a few different vehicles. Instead, this decision is more like buying toys online for a birthday party. You can't try the toys out in advance, but you can look at or read about all the varieties and choose the one you think is best. Don't sweat it, get one or two kits. They all work well enough.

Launch pads for bottle rockets have the complication of a way to connect a bicycle pump. Launch pads also need a way to get a good seal and then a release cable.

In the following pages, first a homemade PVC launcher is shown. Then 6 different commercial launch kits are shown. For each kit, the assembly process is shown, the upper air pressure is listed, and a fun launch is shown. Then the kits are compared side by side on the last page of this chapter. There is no 'right' kit. They are all fun. Some get more pressure, and some are easier to assemble, and some cost less.

Let's compare a water bottle rocket launch pad to a space rocket launch pad.
Launch pads for bottle rockets have the complication of a way to connect a bicycle pump and a way to get a good seal and then a release cable. That of course is nothing compared to the complication of space launch pads.
Space launch pads supply cryogenic liquid fuel and liquid oxygen, and supply clamps to hold the rocket, and supply water to pour on the ground below the hot exhaust, and supply levels or floors for maintenance and access.
Space launch pads must also handle the huge heat of the rocket exhaust. The bottle rocket exhaust of cool water is nothing compared to a huge hot stream of 2000 F degree plasma trying to melt the steel of the launch tower.

'Aquapod' kit

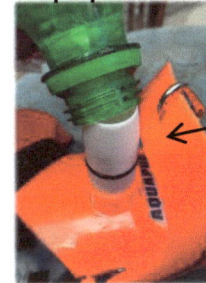

Tight fitting O-ring seal.
A little soap helps slide the bottle over the O-ring.

O-ring seal

bump in PVC pipe

Ties and bump in PVC pipe for seal

Toy rockets can have huge g's, about 20 g's or more, because no astronaut is on board. Space rockets need to keep below a few g's to be safe for people.

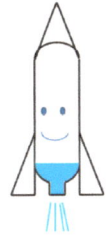

Super-cool Kits, Designs, and Videos of Launches

You can build your own Bottle Rocket launch pad from scratch, or there are many kits to get you setup and going more conveniently. It is much easier to first buy a launch pad kit, which comes complete with launch supplies. The launcher will seal the bottle, pressurize the bottle, and release the bottle.

For the launcher kits, you might need to find a plastic bottle, or get a bicycle pump, but the hard part of getting the launch pad and seal is already done for you. Usually the launcher kits do come with a bottle rocket.

Pumping up the pressure:

Launch kits that take more pressure take more effort to pump up. You might need to invest in a better bicycle pump. The pressure limit may be the bicycle pump, and not the launcher.

A rugged metal launch kit, like the medium nozzle 'Relationshipware StratoLauncher IV', can get past 100 psi, using lots of Teflon tape. The homemade or pre-made PVC launcher with a wide nozzle can get up to 100 psi if done right, and for highest performance the wide nozzle PVC design is more dramatic, and launches with a bang.

Wide nozzles and wider pipes:

High pressure, wide nozzle launches are very exciting, and blast off with a boom with lots of thrust. The launch pad is more challenging to build, because the restraining force is larger and the seal needs to be perfect.

Even launch kits that do not go to very high pressures are fun to use. Low pressure kits, like the 'AquaPod' do the job and launch the bottle. Dissolving baking soda kits create a surprise when the launch happens. There is so much more air drag at higher pressures and launch speeds, that a low pressure launch goes up to a similar height as a wide nozzle, high pressure launch.

Narrow nozzles and narrower pipes:

High pressure kits don't need to start out at faster rocket speeds. The narrow-nozzle 'Antigravity Rocket' is a good kit, no PVC tubes required. It is awesome just to see a bottle rocket blast off like a regular space rocket, with slow and steady acceleration from the slim stream of water out of the narrow nozzle. There is so much more air drag at faster launch speeds, that narrow-nozzle lower-thrust launches go up to similar heights as wide nozzle, high pressure launches, despite wasting energy by raising the water higher.

Launch pad choice:

The launch pad choice is yours: Start from scratch at a hardware store with PVC pipes, or buy one of many pre-made kits. If you want to make your own homemade launch pad, be prepared to spend a lot of time buying PVC piping and experimenting with trial and error to get a reliable air-tight and water-tight seal.

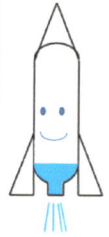

"Rockets accelerating at 1 to 3 g's is a good sweet spot for human astronauts, without feeling like a bug splat on a windshield."

Homemade launch tube

Homemade launcher:
Gather your materials:
- PVC pipe,
- bicycle nozzle,
- garbage ties,
- glue,
- saw,
- wood

Homemade parts, not a kit: PVC parts, where you make a bump or bulge by heating the plastic

Pre-made launch tube

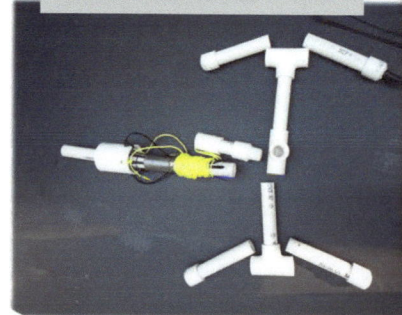

Fully assembled

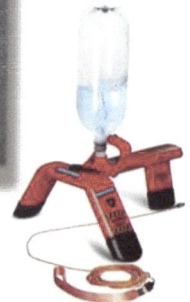

One kit by PVC Creations: User-assembly with pre-made bump and pre-fabricated parts

'Aquapod' kit, already assembled

Toy rockets can have huge g's, about 20 g's or more, because no astronaut is on board. Space rockets need to keep below a few g's to be safe for people.

Launcher Approaches: The Basic Idea with Cork or Clamp

Bottle rockets are just a pressurized bottle pushing out water. To pressurize the bottle, some bicycle pumps can get up to 100 psi, which is a safe pressure for plastic bottles.

Sealing in that pressure without leaks is the challenge. The approach of a seal with a bump or o-ring gets the highest pressure, while just pushing a bung into the bottleneck can only have less pressure because the bung easily pops out.

"Just need a tight seal, so rocket can build up some pressure before self-launching."

Launch tubes, with bump or cork seal

Air pressure

Water

Pull down outer PVC pipe, holding garbage ties against bottle rim, to release rocket.

Pull down rope

Seal: use flush bump in PVC pipe, or use rubber cork with hole for smaller PVC pipe. Rocket is clamped down.

Technique 1: PVC launch tubes, and you have control over the release pressure because garbage ties are holding the rocket down against a PVC bump until released by pulling the rope.
The PVC tube can also use a rubber cork around the PVC tube instead of a bump.
See techniques 1-4 in Appendix C

Kit: Soda Bottle Rocket Launcher by PVC Creations.

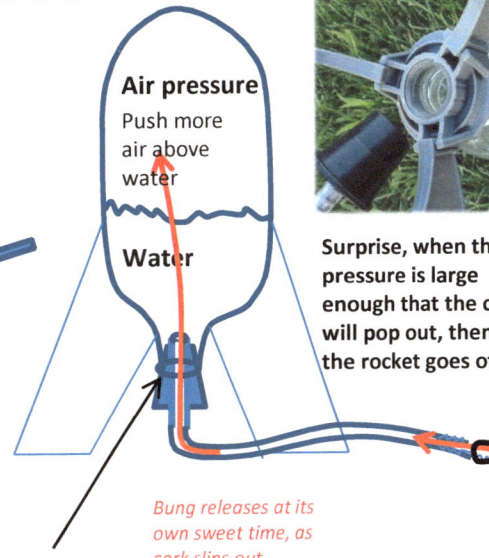

Cork seal without a tube

Air pressure
Push more air above water

Water

Bung releases at its own sweet time, as cork slips out.

Bicycle hand pump

Push air into tube and bottle using bicycle pump.

Surprise, when the pressure is large enough that the cork will pop out, then the rocket goes off.

Seal: use rubber cork or 'bung' with hole and hose in middle. Surprise, when the pressure is large enough that the cork will pop out, then the rocket goes off.

Technique 2: Pressure fit cork, no piping. Rocket launches whenever the cork slips out. Wide corks pop out at lower pressure than narrow corks.
See technique 5 in Appendix C

Kit: Water Rocket by Science in Action

Cork seal and baking soda

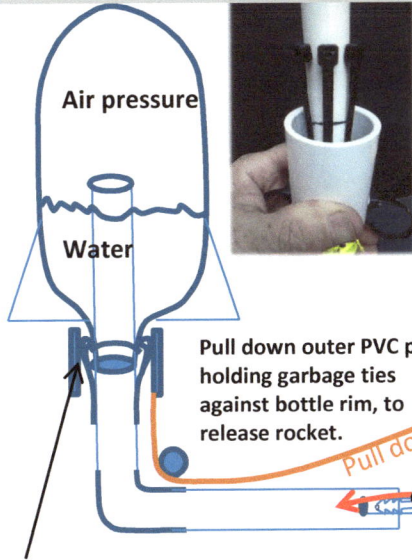

Air pressure

Water + Baking soda

Bung releases at its own sweet time, as cork slips out.

Seal: use rubber cork. Surprise, when the pressure is large enough that the cork will pop out, then the rocket goes off.

Technique 3: Pressure fit cork with baking soda pressure.

Kit: Rocket Science Kit by S.T.E.M. Fun Science Lab.

How to Launch with Water Fuel:
1. Pour water into bottle
2. Tilt launch tube and push into bottle
3. Set on ground, and seal the nozzle
4. Pump up bottle pressure, up to 100 psi (pounds per square inch)
5. Pull on string to release ties and rocket.

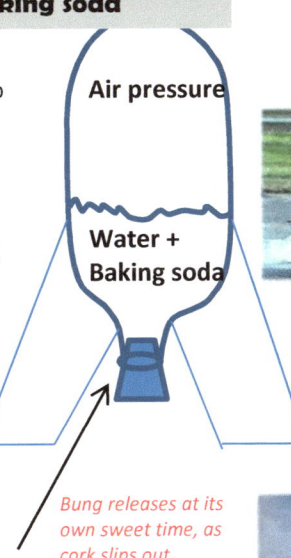

it's not **rocket surgery**

See Appendix C for building homemade launch tubes and seals

A bump or O-ring seal (technique 1) and use a bicycle pump will get up to 100 psi into the bottle. A simple bung (technique 2) gets to about 50 psi with larger bottlenecks and strong insertion pressure. The pressure pushes the bung and water out during launch. Dissolved Mentos pills (technique 3) creates it own gas pressure.

The 'Hard Way' Do-It-Yourself Pad:
Homemade Launch Pad with PVC pipe and Bike Pump Nozzle

How to pressurize the bottle? You need a launch pad to get the air pressure into the bottle.

A homemade launch pad is not so trivial to build, and is a hard way to introduce yourself to bottle rockets. You need to buy some plastic PVC pipes and glue the pipes together, and know how to use heat to make the bump seal or drill holes in rubber stoppers.

For a much easier way to get started go and buy a kit, from 'PVC Creations', 'Anti-gravity', or 'Relationshipware', and the launch pad comes with it.

"Yes, this homemade kluge PVC launcher is the best my dad could do. Let's see if you can do better."

Bike pump stems for launcher

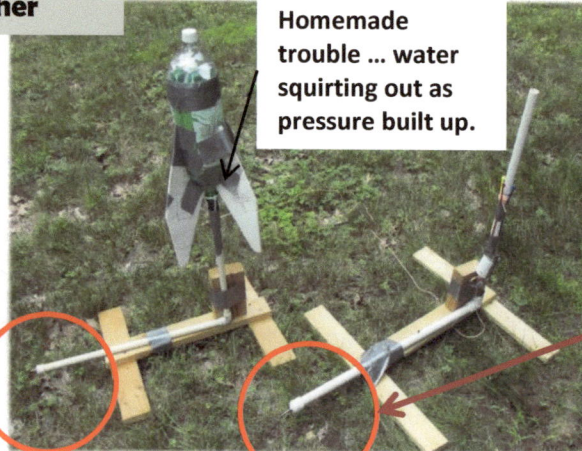

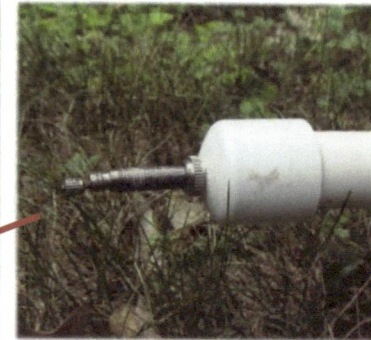

Homemade trouble ... water squirting out as pressure built up.

Two homemade bottle launchers with bicycle stems

Lower pressure car tire stem, for less than 100 psi:
- This stem typically works fine for water bottle rockets, and is easier to connect.
- Used even for launch kits you can buy.
- Automotive stores have this, called a 'tire valve stem'.

Air stem

Higher pressure bike nozzle, for greater than 80 psi:
- Bike stores have this.
- Bicycle pumps to fit these nozzles are less common.

Warning: Stand Back

Space rockets exploding on launch: Water rockets are not as dangerous as chemical rockets, but the plastic bottle can explode with over-pressure or the bottle rocket can land on your head.

Home built launchers take time, planning, and a few trips to the hardware store. Or you can buy a water rocket kit with launcher. Yes, commercial launchers exist.

The 'Hard Way' Homemade Water Launcher

This video sequence shows one of the first launches of this bottle rocket.
The fins were still intact, and the rocket behaves by going straight up.
However, after a few launches, the fins got wet and slightly bent, and the rocket veered off to the side or did flips, as shown in a few pages (Chapters 7 and 8 under 'top weight, bent fins, and stability', or under less stable with bottom weight).

"It is good to be young, dry, and undamaged. Rockets also work better that way."

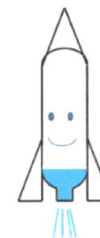

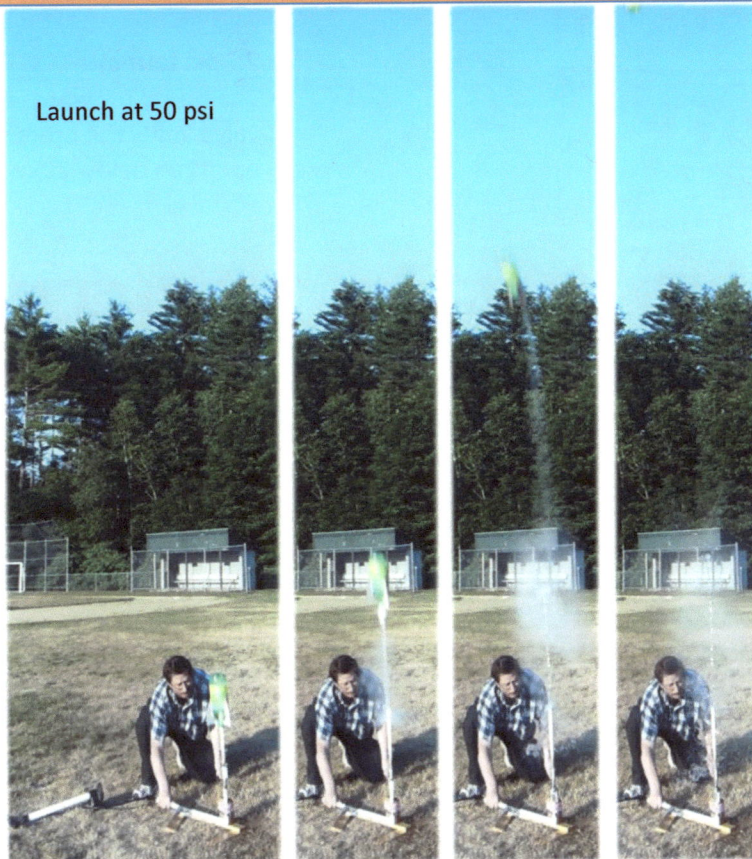

Launch at 50 psi

Frame 2   Frame 3   Frame 4   Frame 5

1/32 seconds per frame.

Hand held launch: The bottle rocket with a wide nozzle was held down by hand against a rubber cork and pumped up to 50 psi. The pressurized bottle was released by hand.
The force on the hand is about 50 lbs. Force equals pressure times area, so a pressure of 50 psi and an area of 1 square inch give 50 lbs force. That is one heavy force to hold back.

At a lower pressure 50 psi, the time to eject the water exhaust takes a little longer, compared to higher pressure 100 psi, but the water is still all ejected within 10 feet.

Fun Facts Extreme designs with high pressure water

Most people don't go to this high pressure extreme, at 500 psi with PVC pipes, but you could. The next level of extreme pressure would be Knievel's super heated steam rocket, at greater than 1000 psi.

Canyon jump: Super heated steam, at greater than 1000 psi

Professional water rocket: super high pressure at 500 psi, and thinner rocket

Fun Facts 'Arrow Principle': Fins and straight flights

Darts with huge fins

Huge fins far behind the heavy pointy front tip allow the dart to be somewhat stable even though the dart is traveling rather slow. The fins help keep the pointy top facing forward.

Fireworks with dragging stick for straight flights

The stick behind the front weight provides air drag and flight stability, by moving the center of pressure and drag force to behind the center of mass, just like having fins.

Kit 1 and 2: Low Pressure 'Easy Way' Launcher Kits

These two kits are perfect just 'to get your feet wet', with less air pressure and easy launch kits. One kit just squeezes a rubber bung into a bottleneck, and the other kit slides the bottleneck over an o-ring. These two kits are pre-assembled in the box.

Although the bottles have a wide nozzle, there is only moderate flow due to the low pressure, so the thrust is moderate.

Rubber 'bung' with wide nozzle

Simple bung bottle rocket with wide nozzle, from 'Science in Action'

Push in bung

'Science in Action' simple bung bottle rocket

20 psi launch:
The bung self-releases at lower pressure, popping out of the bottleneck opening.

Bung releases at 20 psi from air pressure: the rocket only flew to 20 to 30 feet high.

This bung launcher is perfect for young kids.

The commercial water rocket is very easy to set up, from 'Science in Action'. Just squeeze the bung into the bottleneck and pump. The bung releases at around 20 psi, so the rocket only goes up 50 feet. The rocket goes straight up because the fins are provided and very well constructed.

There is a surprise when the bottle launches, whenever the bung squeezes out. The bung releases at a low pressure because it is wide, and there is a lot of air pressure and force on the bung's top surface, given by Pressure*Area, with no garbage ties holding it in.

O-ring seal with wide nozzle

'Aquapod' kit

O-ring seal

Launch check, holding bottle down.

Tight fitting O-ring seal.
A little soap helps slide the bottle over the O-ring.

Lever check keeps the bottle from flying off the O-ring seal, until the rope is pulled.

Pull string to launch bottle

25 psi launch:
The safety pressure release valve limits the pressure and opens up below 30 psi.

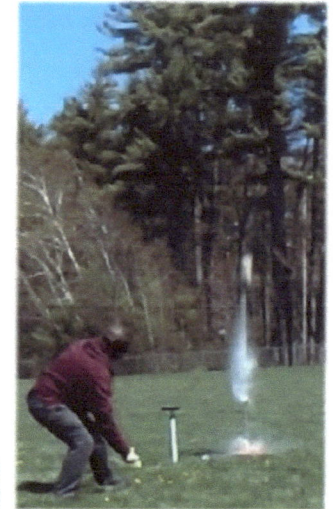

25 psi launch

Pull string to release clamp.
Rocket naturally slides over the o-ring seal and launches from the pressure.

Here are introductory launch pads for sale, which quickly and easily let you start launching water rockets with low pressure.

Kit 3: High Pressure 'Easy Way': Pre-fabricated PVC Kit

This PVC launch kit is a professional version of the homemade PVC launcher with a bump seal, made by heating and compressing the PVC pipe. The bump is pre-made and the garbage ties are already placed, so there is the guarantee of no leaks and high pressure.

The wide nozzle and high pressure with large flow causes large thrust. The flow duration or thrust is very short, because the time to drain the bottle is very short.

'Some assembly required, with PVC glue'

Step 1: Layout of the parts

Layout of the parts, from 'PVC Creations'

Bicycle stem

Inside seal of bicycle stem

Step 2: Glue the parts together using PVC glue

Apply PVC glue to both surfaces

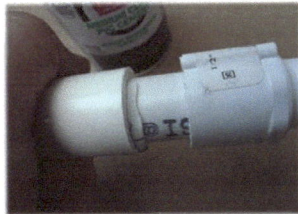

Quickly squeeze the parts together, before glue dries

Finished glued assembly

bump in PVC pipe

Ties and bump in PVC pipe for seal

The fun afterwards

PVC kit with Bump in PVC pipe

PVC kit launch at 80 psi

Launch PVC pipe is easily pumped to 80 psi. The launch goes off with a bang.

This toy launcher name is 'Soda Bottle Water Rocket Launcher Toy Do it Yourself Kit'.

The kit uses a bump in the PVC pipe for the tight seal and a wide nozzle rocket (regular bottleneck diameter) that fits snuggly on the pipe.

This PVC launch pad kit takes some gluing, but this custom kit allows high pressure launches and you only need to go to the hardware store to get the PVC glue. This pre-made kit is a professional version of a homemade launch pad.

Kit 4: High Pressure 'Easy Way' Kit: Narrow Nozzle with Less Thrust, Longer Time for Exhaust

One clever rocket kit gets a narrow stream of water exhaust to lift the rocket, no launch pad required. The narrow stream of water with lower flow and lower thrust is a welcome comparison to the wider nozzle with high flow and 50 g thrust.

The lower thrust means that the water fill should be less, for this lower thrust to lift the weight of the water. The flow lasts longer. Imagine a backyard water hose with a narrow stream from the spray nozzle, or an over-built water gun with a lot of water but still a narrow stream.

The lower thrust means that the rocket air speed is slower at the beginning, so the fins don't have as much air flow and the rocket can steer to the side. Also, for air water rockets, the water weight is all on the bottom, which adds to the instability. See the 'Arrow Principle' later on in chapter 6.

Please recall that lower thrust due to lower flow rate does not mean a smaller final rocket velocity. The low flow rate and low thrust last longer, so the same final rocket velocity is achieved, at least in outer space without gravity.

'Less assembly required, with rubber bands'

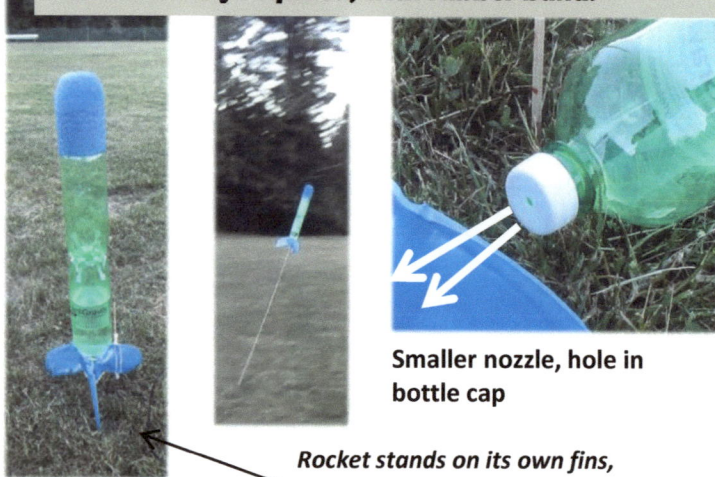

1-stage water rocket: kit with small nozzle from 'Antigravity'

Rocket stands on its own fins, attached with rubber bands. No hardware for a launch pad, other than the air-fill tube. No PVC pipe stand required.

Smaller nozzle, hole in bottle cap

Self sealing tube in nozzle

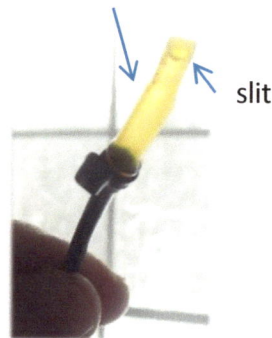

slit

Insert rubber hose in nozzle. The hose has a narrow slit that only opens when the tire pump pressure exceeds the bottle pressure. Otherwise the bottle pressure keeps the slit closed.

"Well, I have a wide nozzle and provide more of a bang!"

"Narrow nozzles work too, and avoid huge air drag."

The fun afterwards

Surprise self launch at about 80 psi when hose pops out on its own.

The water exhaust is a thin stream due to the narrow nozzle.

The air pressure is pumped through the tiny hose, with a 3 mm diameter tiny nozzle.

Launching:
- Assemble rocket with a few rubber bands.
- This is a great launch design for narrow nozzles from 'Antigravity', without PVC pipe. The launch pad does not need large diameter PVC piping!
- The fill hose is pushed inside the nozzle, and has a tiny slit that only opens when the bicycle pump gives pressure. This hose is forced out and releases when pressure is high enough or when the pump is disconnected.

A narrow nozzle can just be a bottle cap with a narrow hole drilled into the center of the cap.
- The smaller flow of water coming out of narrow hole means there is less force, and the rocket can not lift larger water weights.
- The final rocket velocity is the same with a narrow or wide nozzle [+], but the narrow nozzle takes more time with lower acceleration, just like a Space Rocket.

[+] at same pressure and water exhaust velocity.

The thrust from narrow diameter nozzles is more like a space rocket, with lower thrust of 3 g's over a longer time.

Kit 5: 2-Stage Water Rocket, Like the Space Guys

Multi-stage rockets lift satellites into earth orbit. The top stage goes faster with the same amount of fuel.
Multi-stage rockets can be demonstrated even on water bottle rockets, like this 2-stage water rocket from 'Antigravity'. The water is filled separately in each stage, just like a chemical 2-stage rocket.

"Without any heroic high pressure pumping and exhausted arm, my two stage bottles beat the altitude of a single bottle!"

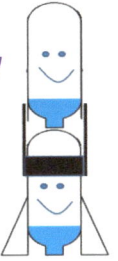

'Less assembly required, with rubber bands'

Bottom 1st stage   Top 2nd stage

Bottom 1st stage nozzle:
The bottom stage has a large nozzle, with larger flow and thrust to lift up both the 1st and 2nd stage.

Tube insert into top 2nd stage nozzle:
The bottom bottle has a tube insert to the top nozzle. This 2nd stage has a smaller nozzle, with smaller thrust to lift up only 2nd stage.
During flight, when pressure in lower stage is exhausted, then the fill tube for the 2nd stage pops out, releasing the 2nd stage, and the water can shoot out the 2nd stage nozzle, propelling the 2nd stage upward.

2-stage rocket, narrow nozzle:
- These kits use narrow diameter nozzles: 3 and 4 mm.
- The rocket does not need a PVC rocket launcher, because the kit has a built in release bulge in the air-fill tube to automatically release at a certain pressure.

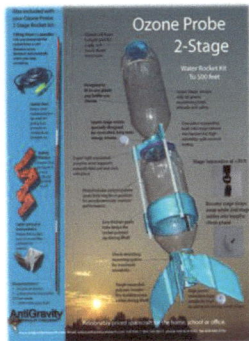

Ozone Probe 2-Stage

Prep for launch:
Air pressure from the 1st stage bubbles up into the 2nd stage, during pumping.

The fun afterwards

2nd stage separates when pressure in 1st stage drops!

1st stage water exhaust

1st stage exhaust runs out

Separation:
2nd stage separates from 1st stage, and has water exhaust from 2nd stage.
Total flight time at 80 psi is 9.5 seconds, longer than a single stage at higher pressure.

Falcon X Space Rocket:
Recent 2 stage rocket to get to low earth orbit and supply the International Space Station.

Falcon X launch

Water rockets can also have 2 stages to get extra height, just like space chemical rockets.

Kit 6: High Pressure 'Easy Way' Kit: Middle Nozzle Diameter

The middle diameter nozzle is unique because a custom made bottle cap is screwed onto the bottle. The cap has a double o-ring seal to allow higher pressures. This launcher kit has a ball bearing release mechanism to release the seal and launch the rocket. These types of o-ring seals are used on air-tool systems.

The medium nozzle with medium flow causes medium thrust. The flow lasts longer, because the time to drain the bottle is longer.

'Some assembly required of metal launch pad'

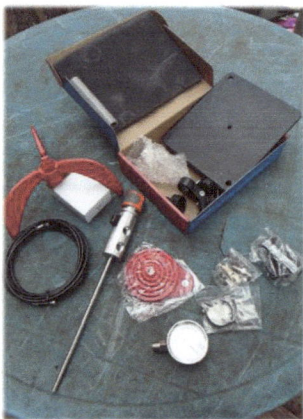

Metal Parts as-is, for 'Relationshipware StratoLauncher'

Metal Parts assembled with wrench and Teflon tape.

Hardware:
The kit contains a 6 mm diameter nozzle, a middle diameter, screwed onto any typical plastic bottle.

The launch kit is metal for strength and reliability, so the only plastic part at the high pressure seal is the cap.

Use plenty of Teflon plumbers tape wrapped around the threads and o-rings to get an air tight seal.

Middle diameter nozzle at 5 mm

Seal pressure:
The beauty of using a reduced diameter nozzle is that the force pushing the bottle out of the launch pad is less, because there is less area. This smaller area places less stress on the water tight o-ring seal. This smaller area also enables the pressure to be higher.

The fun afterwards

80 psi launch
The launcher took the pressure, but the bicycle pump gave out above 80 psi, not the launcher.

Longer thrust and water trail due to middle nozzle

Mounted rocket, getting pumped up

Pull-down cable to release clamp and seal.
Rocket naturally launches from the pressure that the launch pad is holding down, after the clamp is relaxed.

Here is a robust launch pad, made of metal parts, o-ring seals, and tight tolerances to stop water leaking.

Homemade Launch Pads (Hard) or Kit Launch Pads (Easy)

Here is an overview of all the launchers you can build from scratch or buy as a kit, all with different seals and release mechanisms. Think about it. No one hands you just a bottle rocket and walks away, and expects the bottle rocket to just blast off magically without a launcher. The air pressure needs to get into the bottle somehow.

Homemade Launchers:
The hard way, with lots of trips to the hardware store.

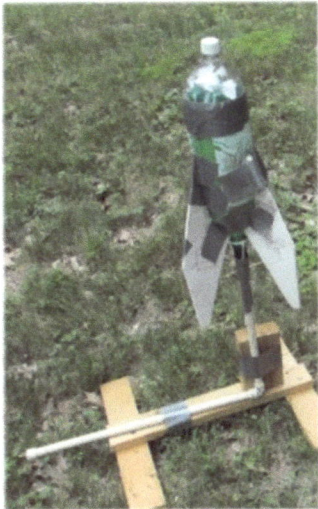

Pre-made Kit Launchers:
The easy way, just follow instructions.

Homemade launcher with PVC pipes and bicycle tire nozzle.
- Glue pipes to get air tight connections and a vertical launch tube with the nozzle
- Heat pipe to get a bump, and get garbage ties in exact position, with a release clamp

See Appendix C for instructions to build homemade launch tubes.

Soda Bottle Water Rocket Launcher, from PVC Creations.
- Prefabricated parts, 10 to 15 minute assembly, toy Do It Yourself Kit.

Science in Action, Bung and fins and bottle
- No assembly, just launch.

'Anti-gravity' Water Bottle Rockets, 1-stage and 2-stage
- Assemble with rubber bands

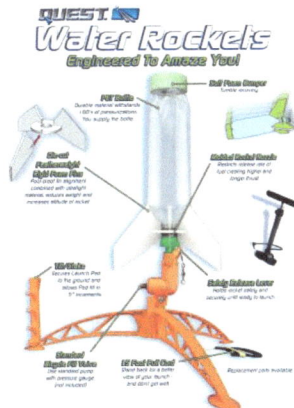

Quest Water Bottle Rocket

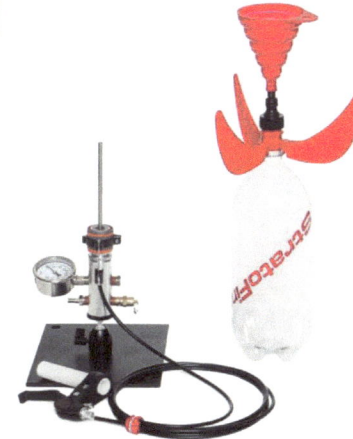

Relationshipware StratoLauncher IV Deluxe Tilting Water Rocket Launcher + StratoFins Kit
- Screw metal parts together with wrench and teflon tape

Launch from chemical reaction, with released gases

AquaPod Water Bottle Rocket
- Snap legs on and launch

Rocket Launcher for Water and Soda Bottles. H- Base/U-Trigger/Pressure-Gauge
Brand: WATER ROCKETS CLUB

You want to do the launch pad the easy way and buy a kit? Good for you, it is more convenient.
You want to make the launch pad the hard way, going to the hardware store many times? Good for you, you are a hardcore engineer.

Types of Seals: Bumps, Corks, Hoses, O-rings

How to seal in all that air pressure and pressurized water without leaking? That is the question all these designs had to answer. Wide nozzles do need to handle a lot of force to hold the rocket down before launch, due to the larger nozzle area: for example, a 100 pounds per square inch (psi) pressure on a 1 square inch tube is 100 lb-force. Garbage ties and o-rings are up to the task (effectively a clamp), while a wide bung is not (no clamp) and slips out to self launch.

Homemade Launchers Seals:
the hard way

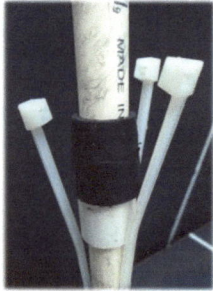

Pre-made Kit Launchers Seals:
the easy way, just follow instructions.

Smaller nozzle, drill hole into bottle cap
'Antigravity' rocket

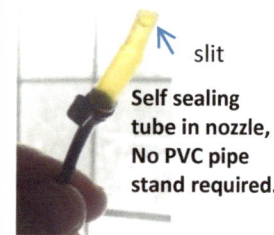

slit

Self sealing tube in nozzle, No PVC pipe stand required.

Rubber cork around PVC pipe,
80 psi.

Rubber corks are nice because the PVC tube does not need to match the bottleneck diameter.

Bump in PVC pipe, using candle heat and pressure,
100 psi.

Bump in PVC pipe,

100 psi.

The smooth bump provides a tight seal with a little water helping to finish the seal.

Expanding rubber hose in nozzle,

80 psi.

The hose will self release when the pressure gets high enough.

Double O-ring seal of medium nozzle,

100 psi.

Medium nozzle diameter puts less stress on the o-ring, so seal can get to the higher pressure.

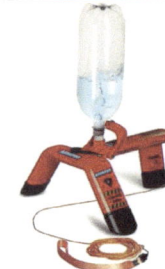

O-ring seal of wide nozzle,

<30 psi.

O-rings are used in car engines, and the Space Shuttle boosters, wherever an air-tight or oil-tight seal is needed.

Bung in wide nozzle,

<30 psi.

Bungs are used in wine bottles and chemistry vials.

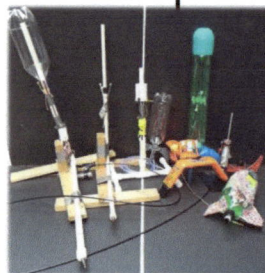

Launchers used in this book

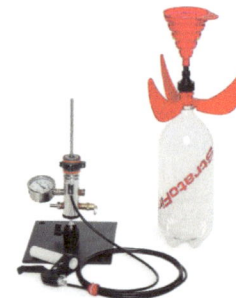

See Appendix A for review comparison of launchers

PVC Bumps, Bungs, O-rings, and Balloon tube inserts: launchers demonstrate different design approaches.

Here are demonstrations and instructions for building bottle rockets powered with air pressure from a bicycle pump. This bottle rocket activity will give disposable carbonated plastic bottles a second life as a bottle rocket, before the plastic from the bottle becomes some recycled jacket, or burnt for power, or simply dumped in a waste dump.

"Water rockets can launch with a bang! Water rockets were one of the highlights of cub scouts, a great outdoor and engineering activity."

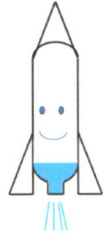

Quick and easy rocket with bottle from store

Larger nozzle

Medium nozzle

Smaller nozzle

Gather people together to build some water bottle rockets. It is great fun to launch different rockets together and see which ones go the highest, go the straightest, and last over many launches. See who can pump up their rocket many times without getting tired.

Very few items are needed to build a bottle rocket. Bottle rockets just need a plastic bottle and some way to pump up the air pressure inside the bottle using a bicycle pump. Of course, you might want to buy a launcher kit.

The bottleneck diameter will determine how much water rushes out, and the thrust. The bottleneck or nozzle diameter controls how much water can squirt out per second. A larger opening will get large flow and thrust, and a smaller opening will get less flow and thrust but for a longer time. The total push on the rocket will be the same, after all the water is gone.

A little top weight will encourage the rocket to fly straight up, even when a fin is a little crooked. Adding a little weight on the top will make the rocket more stable, to shoot up straight even if the fins are slightly bent. Any rocket will be more stable with more weight up top, just like a heavier arrowhead on an arrow.

Air pressure is no stranger to engineering. Here are some air pressure uses. Your car is held up by air pressure in tires. Industrial uses are high pressure tools (pneumatic tools) in machine shops or garages. Household uses are water pressure in your house and flowing water fountains for drinking. Fun entertainment uses are bb guns, and even high-g rides like the 'Screamer' at fairgrounds. Most importantly, air pressure is all around you in the atmosphere, at 14 psi, so that you can breath.

Some larger real rockets use high pressure water, or high pressure hot steam. One of the most famous water rockets, at least in 1970's popular culture, was the steam rocket that Evel Knievel piloted across the Snake River Canyon. Another more practical use of steam is the Poseidon rocket launch using high pressure steam to get out of the ocean from submarines, before igniting the chemical rocket.

Rocket with top weight container and water-resistant fins

Snug fit with PVC bump

The material to build a bottle rocket is all around you, using carbonated bottles, card board, and tape.

Evel Knievel Steam rocket

Poseidon missile

Steam thrust before chemicals ignite

Basic Parts for the Bottle Rocket

Building a plastic bottle rocket is easy. The plastic bottles come at every grocery store. Skip the fins if you want. A nose cone is always optional. The launch pad, however, not the rocket, is the hard part.

For the rocket, you need to cut some cardboard fins and tape the fins to a bottle. The bottle nozzle diameter needs to be equal to the PVC pipe for the launch tube, or equal to a rubber stopper, so the bottle and launcher need to be planned at the same time.

If you skip the fins, accept that the rocket won't fly as straight. If you skip the nose cone, accept that that rocket may get more damaged when it crash lands.

"Rockets can be made from quick cutouts and duct tape. His rocket is ugly but effective ...it was my 9 year old brother's."

Bottle:

A bottle rocket is not as complicated as building a multi-company space rocket. Make a trip to your local stores.

1. Go to grocery store for the bottle. Pick a bottle for carbonated drink, usually with thicker plastic walls. Carbonated drink is at a higher pressure than non-carbonated drink, like pure water. Bring the launch tube PVC pipe with you to test equal diameter bottle necks, if already have the pipe.
2. Go to art or office supply store for the poster board.

Soda bottles at grocery store
Tape and available plastic bottles from grocery stores are great. Pick a bottle designed to store a carbonated drink, which has thicker walls and higher pressure.

Fins:

1. Cut fins out of poster board. The cut should fit the shape of the bottle.
2. Duct tape works well to attach the fins.

Nose cone:

1. Cut top off another bottle
2. Tape this second bottle top to the bottom of the water rocket bottle.

> **The nose cone is not necessary, but the cone should help the following:**
> - Add top weight and fly straight (stability)
> - Reduce air drag
> - Be a 'crumple zone' to help the rocket take the hard collision with the ground.

Brother's rocket

Bottle on homemade rubber seal launcher

Sister's rocket

How hard is it to build any rocket?
Bottle Rockets: Easy. A water bottle rocket is not complex. It is a bottle.
Space Rockets: Hard. The Saturn V 1960's rocket took a team of companies to build, with cold fuel, gyroscopes, steered nozzles, and life support systems.

Saturn V rocket in late 1960s

Saturn V rocket under construction

Build a rocket from a carbonated plastic bottle with stronger thick walls, with poster board fins and tape.

Hardware and Construction for Simple Bottle Rocket

Hardware Items:

Items you need to build a water rocket:

1. Bottle with reasonably thick walls (2 bottles if you want to add a nose cone)
2. Plastic Poster board (plastic does not fall apart when wet, unlike cardboard)
3. Duct tape to hold the fins on (water resistant)
4. Scissors or box cutters
5. Weight on top.

Gather materials

Make sure the bottle fits snug around the tube so air and water don't leak.

Get the right bottle to fit the tube. A 7/8th inch outer diameter tube fits 2 liter bottles.

Snug fit with PVC bump

Construction Steps:

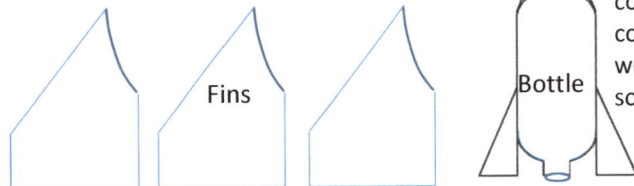

Fins

Bottle

Second bottle cover for nose cone, or top weight like a softball.

"Keep calm, it's only rocket science...

Just get a bottle, some poster board, scissors, and duct tape. Remember that the nozzle size needs to work with your launcher."

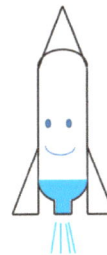

1: Outline:
Draw outline to cut for the fins

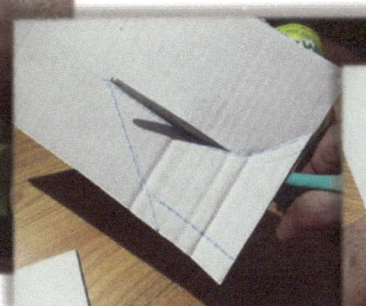

2: Cutting:
Cut fins from poster board, to match the curvature of the bottle.

3: Many fins:
Use the first fin as a mask to trace and cut out the other fins.

4: Tape:
Cut strips of duct tape

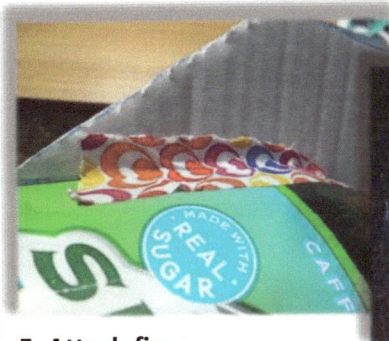

5: Attach fins:
Tape the fins to the bottle. Use duct tape to hold the fins on.

6: Options:
- **Wrap fins in tape:** The fins won't get waterlogged.
- **Top weight:** Put payload compartments on top to add weight, or cut the top or bottom off another bottle and tape it to the top.

Final rocket

Construction is quick to make a simple bottle rocket.

Fill Rocket with Water and Pump to Launch

To launch a water bottle rocket, all you need are a launch pad, a bicycle pump, water, and muscle to pump the air pressure. For water supply, remember to bring a gallon or more of water to your launch site, to have spare water to get many launches.

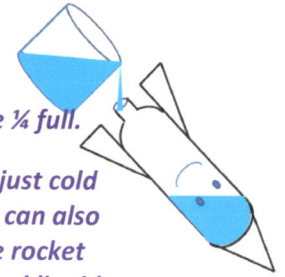

"Just tip me over and fill me ¼ full.

How can rocket exhaust be just cold water? Well, space rockets can also shoot out water if the space rocket combines liquid hydrogen and liquid oxygen as fuel, although their water exhaust is much hotter and faster."

Step 1: Fill water:
Pour water into up-side-down rocket, 20-40%.

Step 2: Insert Launch Tube:
Push PVC pipe into bottle, horizontal, without spilling the water.

Step 3: Aim Rocket:
Tilt the launch pad upright and seal the bottle nozzle so no water leakage.

Stem for bicycle pump on launcher

Step 4: Pump in High Pressure Air:
Attach the bicycle pump and pump up the bottle.

A simple bicycle pump will launch your rocket. Try to pump to 80 psi for fastest exhaust, but even 40 or 50 psi will have impressive launches.

How to Launch with Water Fuel:
1. Pour water into bottle
2. Push launch tube into bottle. Tilt the launch tube so don't spill water.
3. Set on ground, and seal the nozzle
4. Pump up bottle pressure, up to 100 psi (pounds per square inch)
5. Pull on string to release ties and rocket.

Adding water and pumping the pressure is quick to launch a bottle rocket, but only after you have a launch pad.

Here are some tricks for going the straightest and highest, using the 'Arrow Principle'. Anything flying in air – like rocket fireworks celebrations, shooting arrows in your backyard, throwing darts, airplanes flying in air – have the same ideas or tricks to go straight.

How are bottle rockets optimized? They are very similar to the design of an arrow. The arrow gets shot out by the bow.

1. The arrow has no active flight control and depends on the back fins to keep it going in the direction intended.
2. An arrow has a front weight, the arrow head, to help the arrow go straight by moving the center of mass forward.

This chapter explains the 'Arrow Principal', which is nothing more than top weight and bottom fins for stable flight. The principals are based on the fact that things twist around their center of mass. The top weight raises that center of mass, so the bottom fins can have more impact to help stabilize the flight.

Center of mass

Center of pressure below center of mass

Sideways drag force on fins

Arrow Principal: Forward weight and back fins:

Space rockets don't need to think about fins for stable flights because they travel in outer space, without air flow or air drag. Space rockets control their direction by aiming the nozzle. Some top weight is also useful for space rockets, same as an arrow, because the nozzle control will be more responsive – the top weight will provide a stronger torque from the nozzle around a longer lever arm from the farther center of mass.

Thrust and nozzle diameter:

For water bottle rockets, the nozzle diameter is important to control instantaneous thrust. Of course a wider nozzle will have more water flow and more instantaneous thrust. In practice, for any nozzle above 5 mm diameter, the flow rate is already large enough to cause a thrust well above the weight of the rocket, with a water fill ratio appropriate for the thrust value, typically about 25%. A narrower nozzle will have less flow but still the same exhaust velocity. This smaller flow rate but for a longer time will have almost the same final rocket velocity as a faster flow rate for a shorter time.

The nozzle diameter has a direct impact on the amount of water, or fill ratio, in the bottle. With smaller diameter nozzles, below 5 mm diameter, the thrust starts to reduce to just close to the weight of the water, so less water is allowed in the bottle.

Using a small diameter nozzle, the more gradual launch is more like a space rocket launch. For a bottle rocket in regular atmosphere, a more gradual launch means less starting velocity and less overall air drag impact. So a wide nozzle and a narrow nozzle go to nearly the same height. That is just a coincidental outcome of the effects of air drag. Without air drag the wide nozzle would be much higher because the thrust is so much larger than the weight and the impact of the force of gravity is less during the launch.

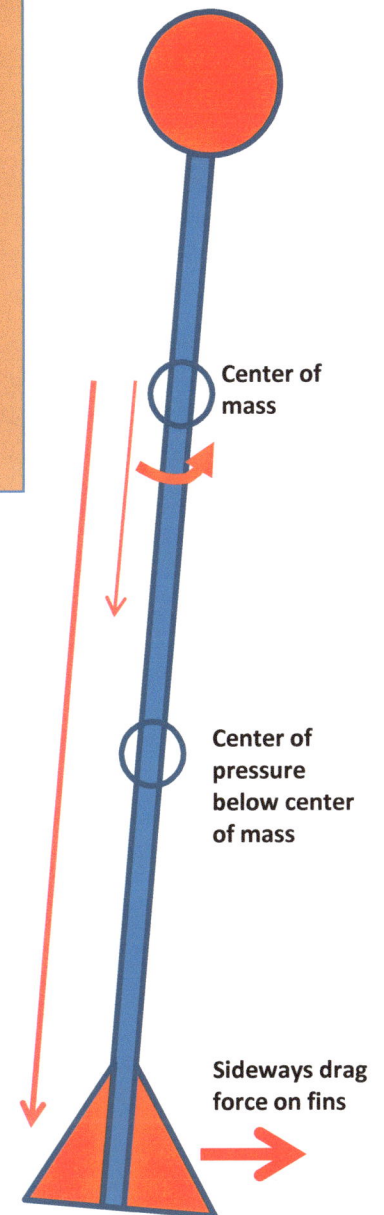

The 'Arrow Principle' rules, with a little top weight and bottom fins

Let's Get More Height and Impress Your Friends

Here are some tricks for going the straightest and highest, using the 'Arrow Principle'. Anything flying in air – like shooting arrows in your backyard, throwing darts, airplanes flying in air – have the same ideas or tricks using center of mass in the front and sideways force on fins in the back.

Use the 'Arrow Principle' with fins and top weight, for better stability and mass and momentum to plow through air drag:

Straight Flights:
- Keep fins far in the back, to keep the rocket going straight. Sideways drag on the fins during a tilt will right the arrow.
- Add a little weight to the top, to keep the rocket going straight as well.

Momentum:
- A little added weight at the top will allow the dry bottle, after the water and compressed air are gone, to still push through the air drag.

Reduce Air Drag:
- Thin rockets do have less air drag. This is especially true if using a wide nozzle, with the huge thrust and high starting speed at the beginning. Faster speeds mean more air drag.

Use the right starting conditions, for pressure and water level:

Pressure:
- Pump pressure between 50 to 100 psi. The rocket height will be proportional to the pressure.

Water volume:
- With the wide nozzle, more water weight can be lifted but don't fill the water more than 40%. We still need the compressed air to have its own volume to power the rocket.
- With the narrow nozzle, less water weight can be lifted so don't fill the water more than 20%. The reduced thrust can not lift extra water. You'll get near the same total final rocket velocity in the end, after all the water is gone, but you still need to lift the weight of the rocket.
- See Appendix J for thrust for different nozzle diameters.

Bottle water rocket

Bottle and arrow use air drag across fins to go straight.

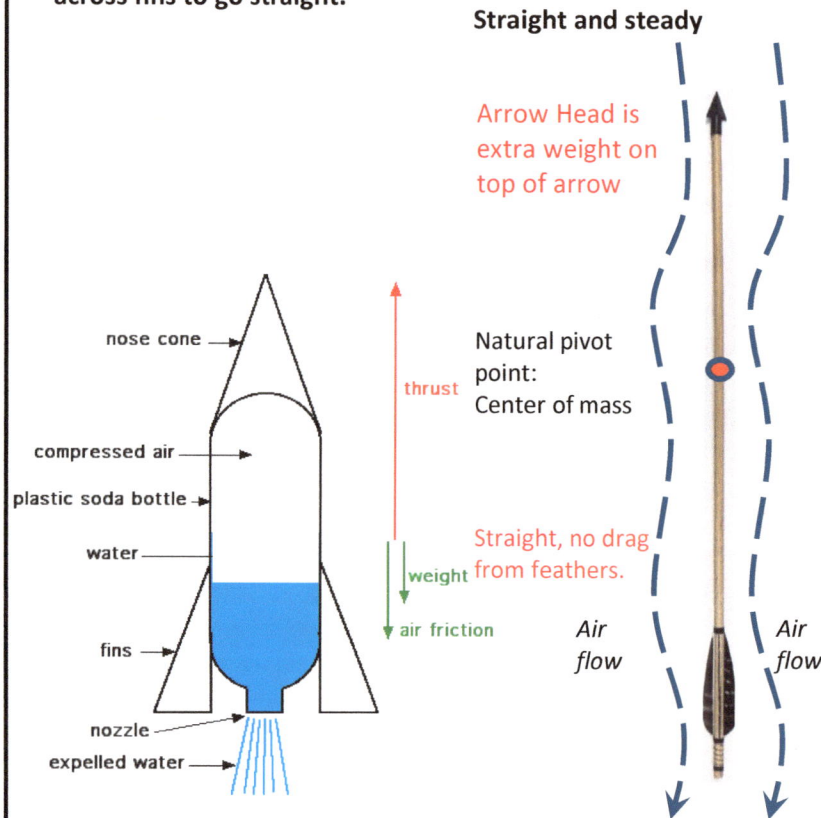

nose cone
compressed air
plastic soda bottle
water
fins
nozzle
expelled water
thrust
weight
air friction

Keep enough air volume: Need air volume, because that is where the energy is from the compressed air.

After thrust is gone, the rocket is basically an arrow, or a fat arrow with air drag.

Straight and steady

Arrow Head is extra weight on top of arrow

Natural pivot point: Center of mass

Straight, no drag from feathers.

Air flow   Air flow

Stability using fins: Tilt, drag from side of feathers rights the arrow (makes the arrow stay in direction of velocity, no wobbling or tumbling)

Space chemical rocket

Space rockets use steered nozzle to balance and turn, not fins.

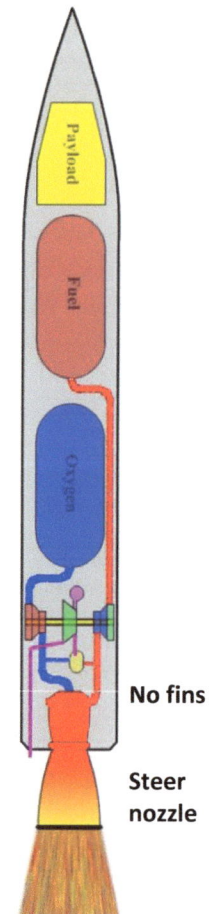

Payload
Fuel
Oxygen
No fins
Steer nozzle

Gimbals aim nozzle and rocket

The 'Arrow Principle' rules, with a little top weight and bottom fins

Make Your Bottle Rocket Go Higher

Air drag is the bane of a fat and light bottle rocket. Stick your hand out the window at highway speeds and you'll know air drag is there.
The most important design tweaks are skinny and long bottles, and a bottle weight heavy enough to plow through air drag. For those of you with an engineering bent, here are design tweaks to get a faster rocket.

"Be skinny and top heavy, to get the most height and stable flight."

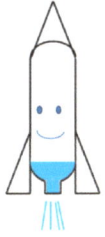

Design tweaks:

1: Use small diameter bottle.
- Small diameter reduces air drag
- A smaller diameter can hold more pressure, for same plastic wall thickness

2: Use a long plastic bottle.
- Long bottle increases volume for air pressure and water fuel.
- Long bottle gets a longer tube and the most tube push. Use long launch tube for more time for air pressure push from tube (before water exhaust starts).

3: Push launch tube high in bottle:
- More volume of air to push the water out.

4: Add top weight: Have empty rocket mass large enough, with a little added top weight for straight flights.
- Put payload compartments on top to add weight, or cut another bottle and tape a cone to the top.
- Air drag does not slow a heavier bottle rocket down as much during coasting, after the bottle rocket reaches full speed and water is gone. A dry weight of about 0.3 pounds (130 gm) dry weight is good if using a 2 liter bottle (see Chapter 8 for top weight, Chapter 9 for air drag, and Appendix G for thrust).
- Use high center of mass, for more stable flight. The sideways drag on the fins when tilted, well below the center of mass, keeps the rocket straight.

5: Cut smooth fins:
- Avoid sharp ends which cause air turbulence and back pressure, so keep smooth curves
- Use 3 fins instead of 4, for less backwards air drag while going straight. When veering off course, then the sideways air drag on the fins will push the back end back of the rocket into going straight.

6: Protect fins from water:
- Wrap fins in tape so the fins don't get waterlogged.

Don't over-fill, and the tube extends above water in the bottle

Rocket with finishing touches:
Top weight compartment, smooth fins, and water protection

During Launch:
Do not overfill the bottle
- ~1/3rd fill, but depends on rocket weight and nozzle diameter. More water just reduces the volume of the high pressure air (that high pressure air is your energy), and requires a larger nozzle to get enough thrust to lift the water easily and go straight.

Use more air pressure to get faster exhaust velocity of the water
- Keep pressure less than 120 psi, or bottle breaks

Nozzle diameter and thrust

Heavy rocket for wide nozzle:
A heavier empty rocket will reduce the impact of air drag (about ½ pound for 12 ounce bottle).

A heavier empty-mass water rocket will still reach close to the same final rocket velocity as the rocket pushes through the air drag, whereas a lighter rocket bottle would have a severely reduced final velocity due to air drag.

Nozzle diameter choice:
1. When use wider nozzle (bottle opening), get large force and faster rocket speed earlier, before gravity has much impact. There is no need to waste energy by lifting fuel and then discharging with slow flow. But air drag is large.
2. When use narrow nozzle, get smaller force and gradual buildup of rocket speed, over 1 second, with less air drag. Fill with less water because the thrust is less, but still get same altitude with less air drag.

Even a wide bottle with no fins is fun to launch. But if you want to go straight, or launch a payload, or go up higher, then you need to think about these design choices – bottle length and width, nozzle diameter and fin size.

Arrow Principle for Straight Flight: Bottom Fins and Top Weight

For stable flight of an un-controlled bottle rocket, like an arrow, the center of mass must be up front and the fin sideways force must be in the back.

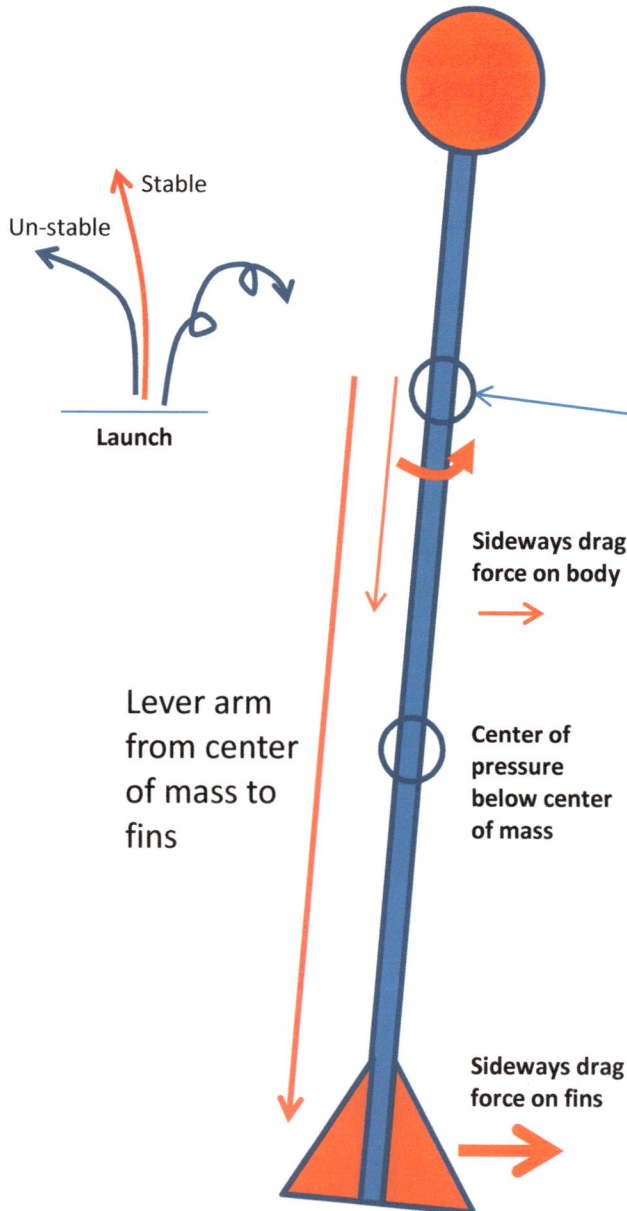

Stable

Un-stable

Launch

More weight on top:
- raises the center of mass for improved stability.
- more weight allows the arrow to plow through air drag easier
- don't add so much weight that the rocket does not fly high anymore.

Center of mass:
- the natural pivot point.
- location is raised using top weight

Sideways drag force on body

Lever arm from center of mass to fins

Center of pressure below center of mass

Longer lever arm to the bottom fins:
- any sideways drag on the fins will straighten out the arrow easier when the center of mass (pivot point) is higher.

Sideways drag force on fins

Bottom Fins:
- drag force sideways because sides of feathers are hitting the wind.

Examples of stable flight using fins in atmosphere

Darts:
Huge fins are needed for stability because darts have slower speed with less wing flow. Also, most of the weight is in the metal tip, which also helps stability.

Fireworks:
Stick provides drag in back like fins, with heavy explosive weight on top.

Arrows:
With fast flight speeds, smaller fins are in back, and a heavy arrowhead is in front

How to get stable vertical flight: straight fins on bottom and a little weight on top

The 'Arrow Principle': Fins on Bottom, and a Little Weight on Top

"My fins help me go straight by causing sideways air drag that counter-acts when I start to tilt."

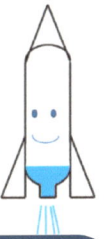

For straight flights, keep fins straight on the bottom, and add a little weight on the top of the bottle rocket, like an arrow. Straight just means go in the direction of launch.

Extra mass at top, to move center of mass toward top, and to resist air drag.

Feathers for sideways drag if rocket not going straight

Tilt Right: Rotate back to straight from sideways drag on fins

Straight and steady

Tilt Left: Rotate back to straight from sideways drag on fins

Lever arm

Arrow Head is extra weight on top of arrow

Natural pivot point: Center of mass

Straight, no sideways drag from feathers.

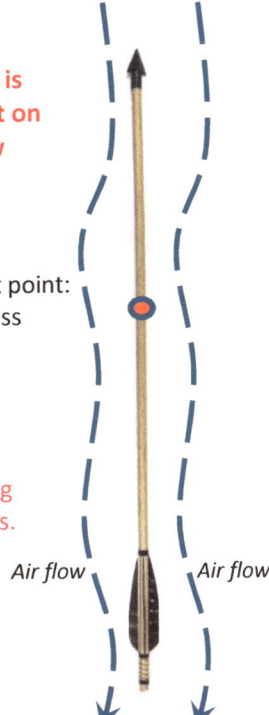

Drag force because sides of feathers are hitting the wind.

Air flow   *Air flow*

Sideways drag force back to alignment

During a tilt, sideways drag from side of feathers rights the arrow (makes the arrow stay in direction of velocity, no wobbling or tumbling)

Sideways drag force back to alignment

- Feathers, or fins, in the back keep the arrow, or rocket, going straight and stable, when extra sideways air drag pushes and rotates the arrow back to straight during a tilt.
- Added weight up front also helps the arrow or rocket going straight, by moving the center of mass closer to front.

The back fins keep the rocket going straight, along with some extra weight on the top.

Stable flight using fins in atmosphere

Stable flight with fins in back: Keep the drag or center of pressure behind the center of mass.

Flight direction

Centre of mass
Centre of pressure

Drag

Un-Stable flight with front wings: Front small wing is an example of deliberately adding lift in front of center of mass, to increase instability and maneuverability.

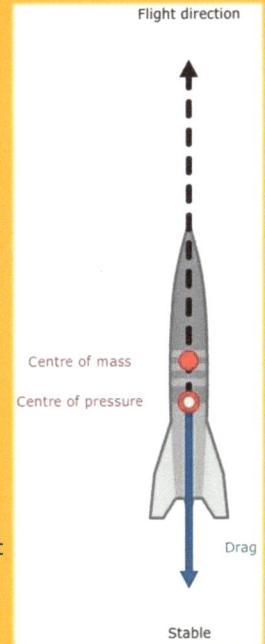

Stable

Center of Pressure is below Center of Mass

Stable flight using steered nozzles in space, without fins

Gyroscope for angle

Nozzles that steer and tilt

Exhaust

Space rockets do not use fins. Fins don't work in the vacuum of space.

Instead space rockets constantly change angle of nozzle to stay upright, as measured using a gyroscope.

'Arrow Principal' Examples

Many things use the arrow principle.
Darts and badminton birdies, and of course arrows, use the arrow principle to go straight.
Airplanes use only the rear wings or fins to go straight, or control the pitch. Airplanes only use part of the arrow principle because airplanes need the center of mass to be close to the center of lift.

A badminton 'birdie' is super light up front, and so needs huge fins in back.

Different speeds and weights require different levels of fin sizes.

Birdies, Darts, and front mass and rear fins:

Badminton birdies and darts do not fly very fast. They have slow air flow, which means less sideways drag on the fins to help keep the direction. Larger fins are needed to keep the stability from the fins, with large enough drag at these slow speeds.

Darts need to go straight right away, even at slow speeds.

Darts are very top heavy and have huge fins in back.

Airplanes and tail fin:

Airplanes need to design for lift as well as stability. The center of lift and the center of mass need to be close to each other, to avoid pitching the airplane up or down.

The tail wing has lots of torque ability because it is far from the center of mass, like fins. The angle of the tail wing and the control flaps on the tail wing keep the airplane from pitching up or down.

Fuel is about 25% of the weight of the airplane, and most of the fuel is in the wings. As the fuel is used, the center of mass of the airplane does not change much because the fuel is over the center of mass.

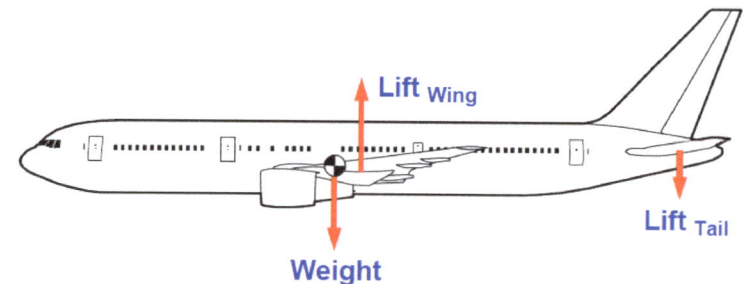

The tail of an airplane provides whatever torque or lift is necessary to stop the plane from tilting due to a difference between the center of lift and center of mass.

Airplanes use the tail wings to control stable flight, just like arrows use the back fins.

Lots of moving things use control surfaces in the back to keep a stable flight.

Aiming a Space Rocket Without Fins: Not the 'Arrow Principle'

Bottle rockets and space rockets have stable flight in two very different ways.

Air Rockets: Air drag and air flow over the fins stabilize the bottle or arrow during flight.

Space Rockets: Spinning gyroscopes are used to measure tilt angle, and pistons and gimballing the engines and nozzles constantly adapt to stabilize the angle. The gyroscope angle remains fixed relative to the launch angle. Watch a space rocket launch video and see the nozzles constantly moving.

"I need to stay upright. Power my pistons and tilt that rocket engine for balance and steering."

Bottle Rockets in air: top weight and fins

Weight up top for stability

Fins on bottom to keep outside 'center of pressure' low.

Self launching water rocket with fins, at 25% fill

Stable flight:
The rocket is stable due to air flow and drag, as long as Center of Pressure is below the Center of Mass. However, this air drag does not exist in the vacuum of space. Space rockets do not use fins, and instead use vectored thrust by re-aiming nozzles.

Location of water weight:
For the water bottle, note that the water on the bottom actually contributes to instability, because the center of mass is low. That low mass is why water bottle rockets can fly sideways right after launch, and need a guide rod to let the rocket build up speed and wind over the fins before let loose in free flight.

Space Rockets in a vacuum: Re-balancing using steering of nozzles

Toy gyroscope: When a spinning disk is placed inside rings, the disk can hold its orientation, and indicate 'up', even as the rocket moves to any direction.

Piston to control orientation of nozzle, with feedback from the gyroscope.

Single engine, aimed with pistons

Space Rockets: Multiple engines to allow more steering control

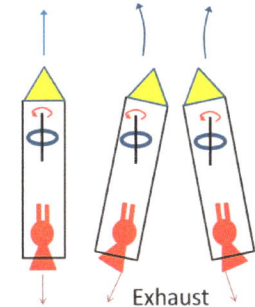

Exhaust

Bottom balancing: Move your hand quickly under a pole or feather to keep it balanced, and stop a fall. Balancing from the bottom is easier when there is a heavier weight on top.

Space Shuttle: Gimbaling engines, with independent movement. Nozzles dynamically change direction with fast updates to re-aim and balance the rocket, using gyroscopes and pistons.

Bottle rockets use fins and air drag to have stable flight. Space rockets, without air flow, use gyroscopes and vectored nozzles to aim the rocket and keep the rocket from falling over.

Water Rocket Prep for Launch, and Forces

How much water to fill to go the highest? To fill or not to fill, to pump high pressure or not to pump high pressure, that is the question!
Water bottle rockets powered by air pressure have their own concerns, different than chemical rockets. Wide nozzles cause huge thrust with huge flow, but for a very short time. Narrow nozzles have less thrust with less flow and so can not lift as much water. For both wide and narrow nozzles, more than half the volume should be pressurized air because that's where the energy is, to get maximum speed and altitude.

"Most any water amount works below 40%, but some water levels work better. ...don't use more stored water than the thrust can lift, especially for the light water stream from a narrow nozzle."

Pumping and Water Setup

Wide Nozzle (like a bullet, bang!)

Air Pump Pressure:
Between 30psi and 120psi:
- If greater than 120psi, hey, the bottle's only thin plastic, and the bottle can rip open. Safety first!
- If less than 30psi, don't you want the rocket to go high? What about fun?

Narrow Nozzle (like NASA rocket, slow and steady)

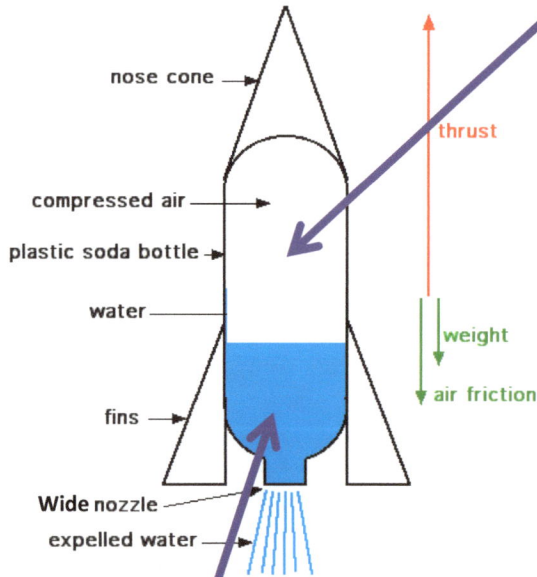

Forces:
Thrust, gravity, air drag

Push Water Out for Thrust:

Gravity Down:
- Force down is Force = mass*acceleration of gravity

nose cone

thrust

compressed air

plastic soda bottle

water

weight

thrust

air friction

weight

fins

Where's the energy?
High pressure air inside bottle.

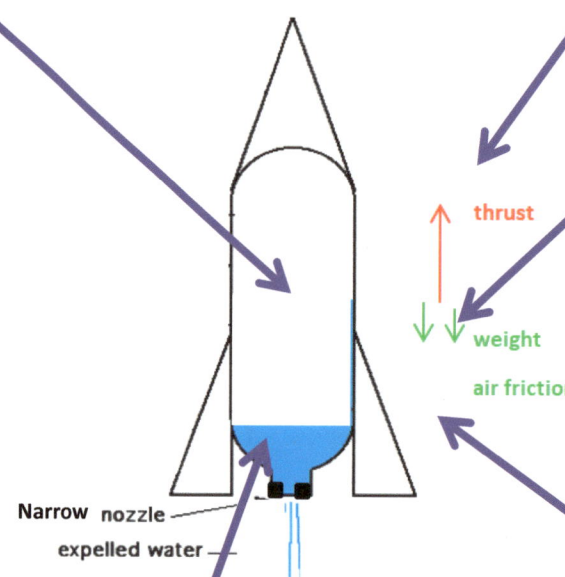

air friction

Air Drag Opposite the Velocity:
- Force of drag due to air is experimentally known to go as velocity squared, so you get penalized with drag much more the faster you go.

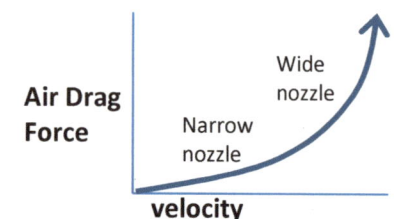

Wide nozzle

expelled water

Narrow nozzle

expelled water

Fill Bottle for Wide Nozzle:
20-40% full of water:
- If bottle has greater than 40% full of water, then there is too little air for pushing.
- If bottle has less than 20% full of water, then too little water is pushed out before the rocket really gets going.

Fill Bottle for Narrow Nozzle:
10-20% full of water:
- There is less thrust with less flow, although for longer time. We use less water because the smaller thrust still needs to lift the water. A larger water weight will cause rocket to tip sideways, with less speed and fin air flow and unstable bottom weight.

Air Drag Force

Wide nozzle

Narrow nozzle

velocity

http://www.ohio.edu/mechanical/thermo/property_tables/gas/adiabatic/rocket/analysis2.html

Pump up the air pressure high, and the water about 20% volume. Release the rocket and let the forces (thrust, gravity, air drag) do their work. A rocket with a wide nozzle, with more instant thrust, can lift more water than a rocket with a narrow nozzle.

Example: Nozzle Size for Space Rockets

Space rockets and water bottle rockets have the same physics about thrust. Some large space programs have used a few very large rocket engines, and some have used many smaller engines. Both get the same total thrust.

Many rocket engines, or an array of engines, are necessary in order to roll and pitch to get the right path to orbit. In contrast, a water bottle rocket does not need to have any particular flight path, and there is no control feedback anyway, so one nozzle will do.

A large nozzle just means the thrust is much larger than the weight of the rocket just using that one rocket engine.

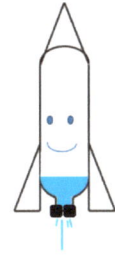

"Space rockets can have a design with only a few large nozzles."

"Space rockets can also have a design with many smaller nozzles."

The rocket launch needs a thrust greater than the weight of the rocket. Space rockets have 3 gs of acceleration and use 2 or more rocket engines, each of which is a medium nozzle compared to the weight of the rocket. This allows the rocket to roll and pitch to get the right path to orbit. Missiles can have 50 gs of acceleration, without people on board, and effectively have wide nozzles.

Patriot missile:
A single engine with a thrust much larger than the weight of the rocket.

SLS rocket, 2020s:
A few engines, each with large flow rate and a few million pounds thrust.

SpaceX, Falcon Heavy, 2010s:
More engines, each with less flow rate but more ability to handle one engine that does not work.

Larger wide nozzle:
large flow and large thrust
Thrust ~ 120 lb-force

Medium nozzle:
medium flow and medium thrust
Thrust ~ 15 lb-force

Smaller narrow nozzle:
small flow and small thrust
Thrust ~ 2.7 lb-force

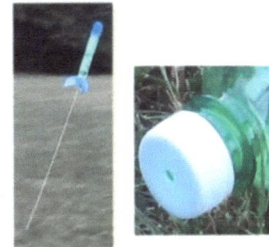

A wide nozzle can just be the open bottle.
More water flow coming out of a wide hole means there is more force, and the rocket can lift larger water weights.

A medium nozzle is a fancier adapter cap on bottle.
The narrower bottle neck has o-ring seals.

A narrow nozzle can just be a bottle cap with a narrow hole drilled into the center.
Less water flow coming out of narrow hole means there is less force, and the rocket can not lift larger water weights.

Even though the thrust from Space Rockets is huge, the mass is also huge, so the acceleration is about 3 g's.

Narrow Nozzle: Less Thrust, Longer Time for Exhaust

What thrust do you want? You can design the water rocket to blast off like a madman with a wide nozzle, or be more gradual with a narrow nozzle. There is more thrust with more water flow using a larger nozzle or opening. The same large opening rule applies to a fire hose, or a chemical rocket. No matter the hole diameter and thrust, the final momentum change of the rocket is the same. Faster flow rate means more magnitude but less time for thrust, for the same amount of fuel.

The rocket launch needs a thrust greater than the weight of the rocket. A half full bottle with water weighs about 2.5 pounds. The large and medium nozzles can lift that weight. The narrow nozzle can not.

Also, thrust equal to weight will just cause hovering, not acceleration.   So a thrust a factor 2 or more greater than the weight is best.

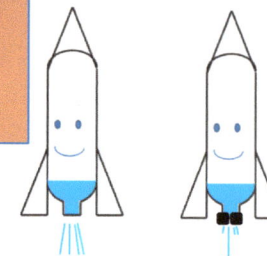

"Well, I have a wide nozzle and provide more of a bang!"

"Narrow nozzles work too, and avoid huge air drag."

Larger wide nozzle: large flow and large thrust	**Medium nozzle:** medium flow and medium thrust	**Smaller narrow nozzle:** small flow and small thrust
Thrust ~ 120 lb-force for 100 psi pressure and 20 mm diameter nozzle.	Thrust ~ 15 lb-force for 100 psi pressure and 7 mm diameter nozzle.	Thrust ~ 2.7 lb-force for 100 psi pressure and 3 mm diameter nozzle.

A wide nozzle can just be the open bottle.
More water flow coming out of a wide hole means there is more force, and the rocket can lift larger water weights.

A medium nozzle is a fancier adapter cap on bottle.
The narrower bottle neck has o-ring seals.

A narrow nozzle can just be a bottle cap with a narrow hole drilled into the center.
Less water flow coming out of narrow hole means there is less force, and the rocket can not lift larger water weights.

Wide nozzle thrust to weight ratio:
- Water (25% fill) and bottle weigh about 3 lbs, so there is 50 times more rocket thrust than gravity, or 49 g's acceleration.

Medium nozzle thrust to weight ratio:
- Water (25% fill) and bottle weigh about 3 lbs, so there is 5 times more rocket thrust than gravity, or 4 g's acceleration.

Narrow nozzle thrust to weight ratio:
- Water (10% fill) and bottle weigh about 1.5 lbs, so there is 2 times more rocket thrust than gravity, or 1 g's acceleration.

Experiment: Put in more total water, and see if rocket tilts or falls to the side during launch. The weak thrust from the narrow nozzle can't push up the heavier water as well.

Launch of narrow nozzle rocket

Narrow diameter nozzles are more like a space rocket, with a lower thrust comparable to twice the weight over a longer time.

"Air can shoot out the nozzle too and cause thrust.
Air can also push water out first, but then the air gets pushed out too which provides thrust as well."

Any bottle rocket you fly, even originally filled with water, will have the final part of the thrust that is just air exhaust. Air thrust works. A bottle rocket without water goes up almost as high as a bottle with water in it. That thrust is because the air has mass, the air is shooting out the bottleneck nozzle with a very high exhaust velocity.

A bottle rocket with water will first of course shoot out the water. But then there is still the remaining pressurized air, which is still at more than $\frac{1}{2}$ the starting bottle pressure.

There are other air rocket blast applications, besides toy bottle rockets, like space walks, CO2 Derby cars, and even aerosol cans:

One life-saving use for air rockets is for astronauts in space. If the astronaut gets loose during a space walk from the space craft, like the International Space Station, then little blasts from compressed air in a backpack will push the astronaut back to the craft.

As a toy, the CO2 derby in Boy Scouts is an air rocket. The exhaust from a high pressure cylinder pushes the car along the ground. The cylinder comes pre-pressurized.   The seal needs to be broken with a pointy object like a nail. No gravity is involved, unlike the Pinewood Derby races. These CO2 cylinders are mostly used for pellet guns, but a CO2 derby car is another use.

An aerosol can for bug spray or hair spray has air exhaust, but that flow rate is small so it is not designed to use its thrust.

Bottle rocket with air only, no water

Spacewalk: SAFER jetpack in space, no tether.

Double barrel CO2 rockets for derby car

Air bottle rockets are great toys.   Just push some high pressure air into some container, like a bottle or balloon, and release the air pressure.
Air is the exhaust, and air has mass just like any water exhaust has mass.

Yes, Bottles Without Water, Just Air, Launch Too

Here is an air rocket, in a professional, museum grade rocket demonstration. At a push of a button, high pressure air fills the up-side-down bottle to 90 psi, and boom! The bottle is released, and shoots up a tube using air exhaust.

"Now that's a great demo! Even the people at Kennedy Space center think bottle rockets are cool."

tube

Air Rocket Exhibit at Museum at Kennedy Space Center, FL

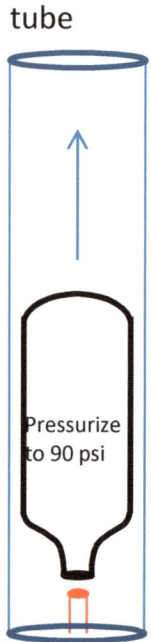

Pressurize to 90 psi

Bottle getting pressurized

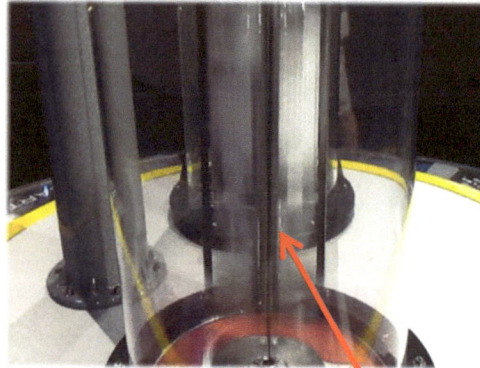

Fast blurred launch

Museum hands-on activity demo, an air-only rocket

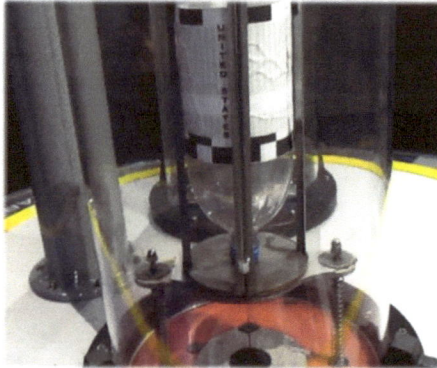

Pressure gauge: 90 psi

The rocket must be extremely light to go up high, because air is ejected and empty very quickly, so this type of rocket can not take much of a payload.
To relate to chemical fuels, a space command module is too heavy and can not be launched into orbit using compressed air.

Push Plunger

Air-Powered
SLAP ROCKET

Push the accordion to launch

Fill Balloon

An untied balloon is also an air rocket! The balloon just is so floppy that it does not like to go in one direction.

Balloon on a line

Stomp Rocket

Get pressure blast by jumping on pump.

Quick jumps to launch

An air-only rocket shoots out the air, no water, and that works fine while the air is shooting out. Think toy Nerf rockets. An air-only rocket is the extreme limit of no water in the water rocket.

The museum exhibit shown above keeps the bottle rocket in a tube, because this is an indoor exhibit.
- This in-door exhibit could pump a little vacuum above the air rocket to let the bottle go higher, but that would be cheating to get into outer space. For some applications, a vacuum is possible and would remove air drag, like on future proposed people-movers in cities where people move in sealed pods in vacuum tubes.

Air bottle rockets are great toys. Just push some high pressure air into some container, like a bottle or balloon, and release the air pressure.

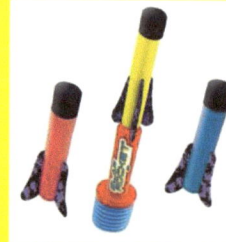

Top Weight, Bent Fins, and Stability: Air Rocket Launch, Flips at 40psi

Don't believe that air pressure without water still launches the rocket? Don't believe that top weight helps? Here's a demonstration. Air exhaust can even lift a softball on top, and the rocket flight is very straight.

"Extra weight on top allows a straight flight, even though one of the fins was bent."

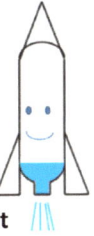

Flight is less stable with bottom weight:
- Air rocket without added weight (no payload, no water)
- Rocket flips around, especially with slightly bent fins.

Flight is more stable with top weight:
- Air rocket with added softball weight on top
- Rocket stays vertical, even with slightly bent fins.

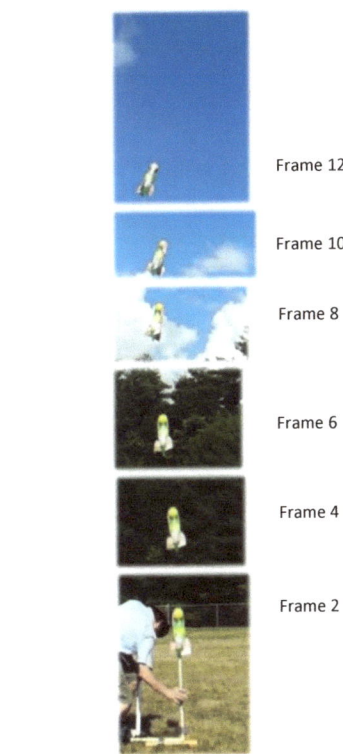

What happens with bottom weight? Is the flight stable?

The rocket blasts off quickly with air and the light rocket.

However, the rocket is unstable because the center of mass is down closer to the fins, and the fins are not perfectly aligned.

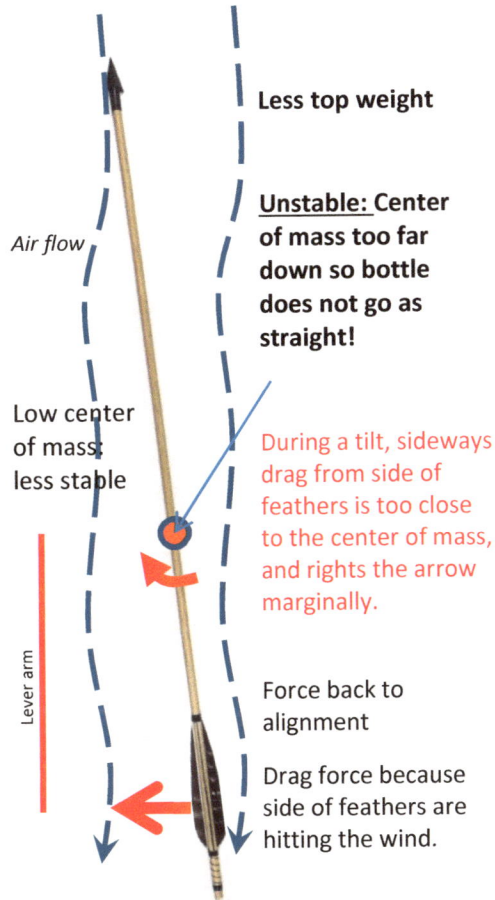

Frame 8

Frame 6

Frame 4

Frame 2

Less top weight

Unstable: Center of mass too far down so bottle does not go as straight!

Air flow

During a tilt, sideways drag from side of feathers is too close to the center of mass, and rights the arrow marginally.

Low center of mass: less stable

Lever arm

Force back to alignment

Drag force because side of feathers are hitting the wind.

No top weight: Rocket flops around with slightly bent fins.

No top weight, ready to launch.

What happens with top weight? Is the flight stable?

The rocket blasts off more slowly with the softball on top. This is expected because there is more mass to lift.

The rocket stays vertical, even with a bent fin, with the added softball weight on top. This is because the rocket is pivoting about the heavy softball on top, and the drag on the fins can re-pivot the body in the direction of travel.

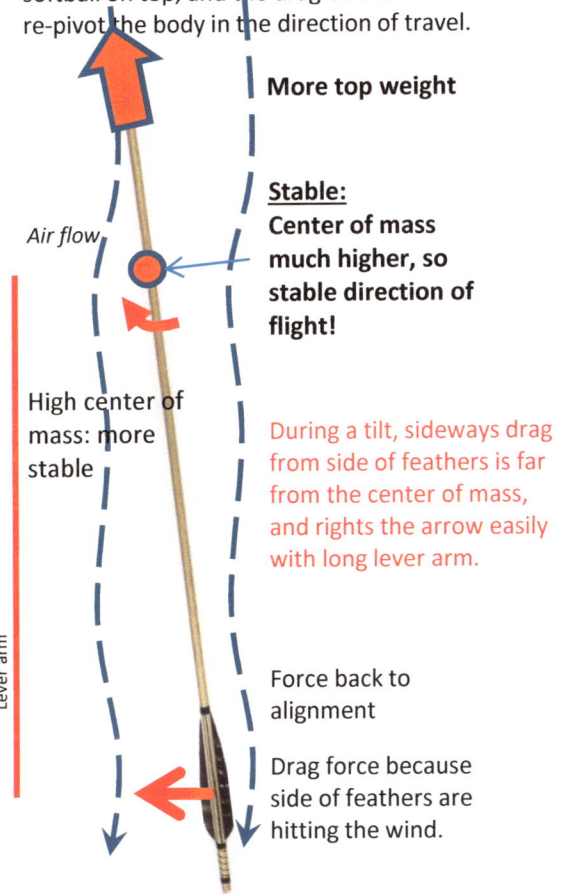

Frame 12

Frame 10

Frame 8

Frame 6

Frame 4

Frame 2

Top weight: Rocket goes straight, even with slightly bent fins. A heavy softball was the added weight at top of rocket to provide stability.

Softball on top of bottle rocket

Top weight of softball, ready to launch.

More top weight

Stable: Center of mass much higher, so stable direction of flight!

Air flow

During a tilt, sideways drag from side of feathers is far from the center of mass, and rights the arrow easily with long lever arm.

High center of mass: more stable

Lever arm

Force back to alignment

Drag force because side of feathers are hitting the wind.

This air rocket demonstrates how a heavier weight on top raises the center of mass pivot point, and allows the fins to right the rocket.

Example: Air Exhaust and Astronaut Mobility

The idea for a compressed gas bottle rocket is used in space for untethered flight. There is just gas exhaust with no chemical reaction. This exhaust gas thrust is very safe and reliable. The rocket thrust is very small, but we just need a gentle way to get astronauts back to the space station.

Compressed gas is used to fly untethered around the International Space Station. This SAFER jetpack goes over the life support Extravehicular Mobility Unit. SAFER stands for Simplified Aid For EVA Rescue.

SAFER is designed as a precaution is case tethers, safety grips, and the robot arm can not help the crew member get back inside the space station. The SAFER exhaust enables self rescue for the astronaut.

SAFER can provide a total change in velocity of 3 m/s.

On the SAFER, the hand controller provided six degrees-of-freedom (DOF) maneuvering via 24 gaseous-nitrogen (GN2) thrusters. Vehicle weight is 85 pounds. The GN2 is stored in four cylindrical tanks, each charged to 3250 psi. Total fuel capacity is 3 pounds which is sufficient to change the vehicle velocity up to approximately 10 feet/second (ΔV).

1. **Is compressed gas really used for astronauts?**
   - N2 exhaust comes from 4 high pressure cylinders.
   - Compressed gas is cheap and reliable, as opposed to lighting a chemical reaction.

2. **Have chemical rocket jet packs been used in space as well?**
   - Yes, some more expensive jet packs using chemical propulsion could go much faster. But tethers and slow astronaut speeds around the space craft make the higher speeds unnecessary.
   - The names are 'Astronaut Maneuvering Unit AMU' and 'Manned Maneuvering Unit MMU'.

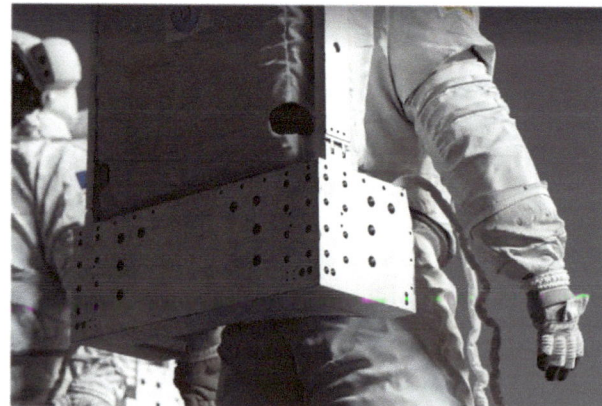

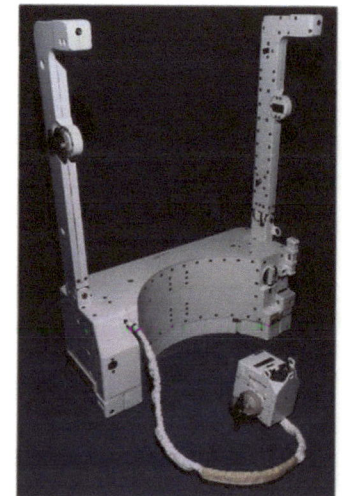

Spacewalk: SAFER jetpack in space, no tether.

Here is the spacewalk in orbit, using the SAFER jetpack, which can blast compressed nitrogen to steer back to the space station.

SAFER jetpack which goes over the life support suite.

Compressed air is shot out of 24 gaseous-nitrogen thrusters to move astronauts around the ISS in an emergency separation.

Example: Air Exhaust and Satellite Attitude Control

Besides astronaut mobility, the idea for a compressed gas bottle rocket is also used in space to keep the orientation or angle of the satellite. Compressed gas is useful and reliable.

Compressed gas is used to stabilize the angle or attitude of satellites. The antennas need to point at the same place on the Earth, so the satellite needs to spin at the same rate it orbits the Earth. Telescopes for the stars and universe need to point at the same direction in space, so those satellites need to not spin relative to other stars, even as the satellite orbits the Earth.

1. **Is air really an exhaust for a bottle rocket, or any rocket?**
   - For satellites, compressed gas can provide the small angle and rotation adjustments to keep a satellite either pointing at a star or rotating to point to a constant angle to the ground.

2. **Can compressed gas keep the satellite in orbit?**
   - No, small chemical rockets will burn to keep the satellite in orbit, as opposed to just controlling the rotation of the satellite. Chemical rockets have a lot more thrust and energy.

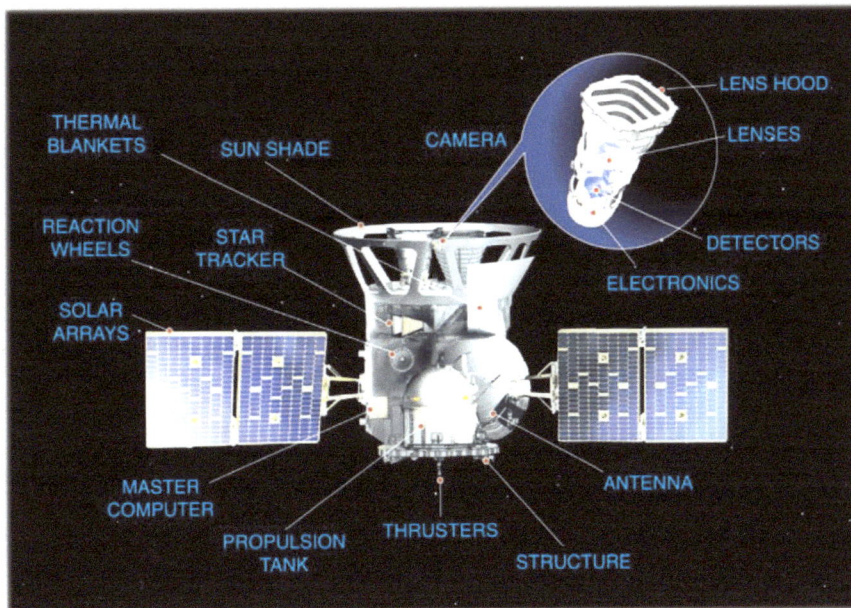

Example: Transiting Exoplanet Surveying Satellite (TESS):
- Search for periodic dimming of stars, indicating a passing planet

The thrusters eject compressed gas from high pressure gas cylinders.
A star tracker will provide feedback to determine if the satellite needs adjustments to keep pointing at the same direction relative to the stars.

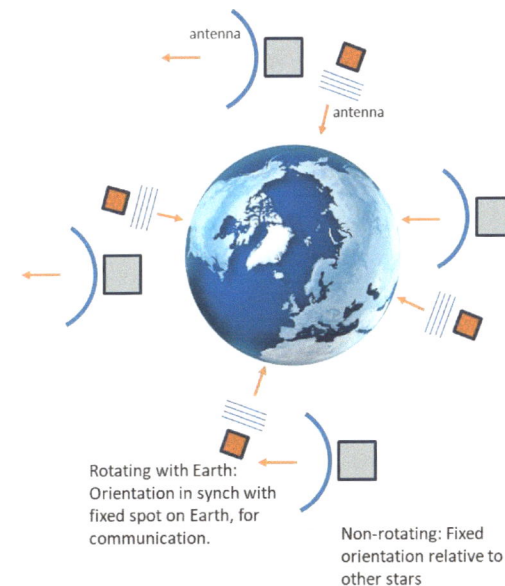

Satellites can either rotate about their axis to keep pointing at Earth, usually with a 24 hour period, or satellites can not rotate and stare at a distant star.

Compressed air exhaust is used to control the attitude of satellites, for antennas point at the ground and telescopes pointing into space.

Example: High Pressure CO2 Powered Rocket Cars, on Straight-Away

A compressed gas is already made in a cylinder, called a CO2 cylinder. The cylinder is usually used for pellet guns, but we can also use the exhaust to power a toy car.
You can add rocket power to a Pinewood Derby, just for kicks, when you graduate from Cub Scouts into the Boy Scouts. Now the competition is called CO2 rocket cars.

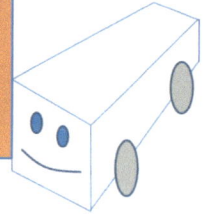

Hole for CO2 cylinder, typically used for pellet guns.

CO2 car with only one cartridge.

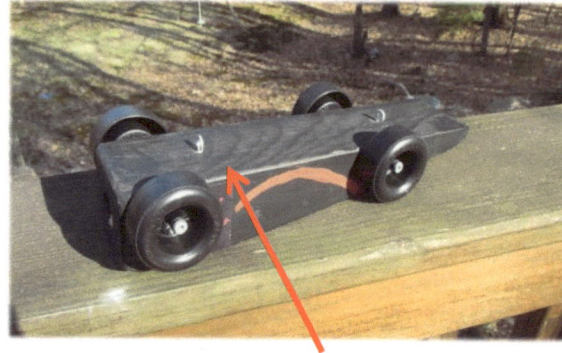

Eye screws on bottom for guide rope.

"These are Boy Scout races, with rocket power. Older scouts graduate to powered compressed gas CO2 cars on straightaways (flat floors), no gravity.
This exhaust thrust makes it a rocket."

High pressure gas CO2 powered cars use exhaust gas like a rocket for thrust. These cars do not use gravity, and travel on a flat ground along a guide rope.
For these rocket powered cars we want the least weight, so the CO2 exhaust will push on less mass.   This is not a gravity car.

* From internet

Single CO2 cartridge or cylinder design

Ways to make a quick hole to start race?
- Hit puncturing pins with a hammer
- A bumper resting against the starting gate releases a mouse trap which punctures the CO2 cartridges.
- If on the sloped Pinewood Derby track, you could make a parachute flap contraption which releases a mouse trap, to trigger the puncturing pins to pierce the seal of a CO2 cartridge.

Double barrel thrust

The older Boy Scout cars can go faster than gravity, using CO2 rocket power!

High Pressure CO2 Gas Rocket Force

"Yeah, I got a jet pack on!"

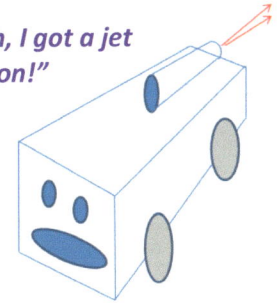

These double barrel CO2 rockets are ready to fire, for this Boy Scout activity! Just hammer them both open at the same time, and get double the force and acceleration.

Unlike the gravity powered PD cars, the mass of the car has a big impact compared to the fixed thrust of the CO2 exhaust. The thrust is not from gravity but from a rocket engine, so less mass means more acceleration and speed. Gravity force is proportional to mass, and rocket thrust is not. Rocket thrust is proportional to the exhaust rate.

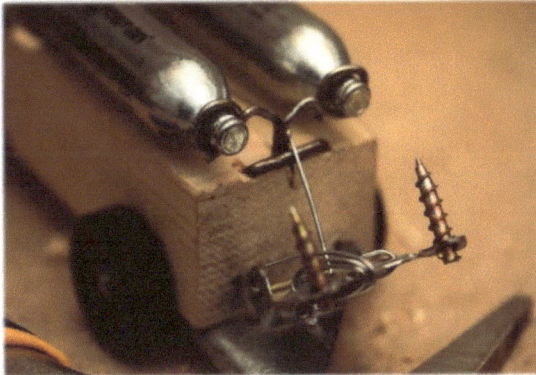

Gas pressure CO2 canisters ready for puncture

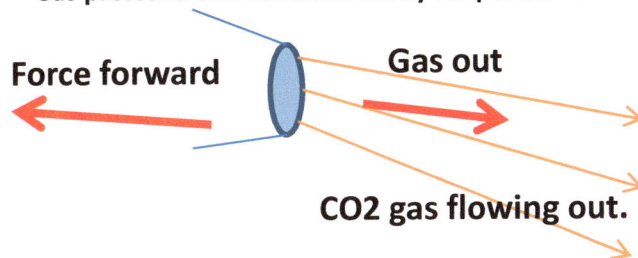

Force forward   **Gas out**

CO2 gas flowing out.

The CO2 gas should accelerate the car just like a rocket engine, even on level ground with about a 2 pound-force. Assuming the gas does not empty out, a larger diameter hole will cause the car to go faster right from the beginning.

The gas pressure inside a CO2 canister is about 800 psi. If you puncture a 0.04inch (1mm) diameter hole, the ballpark force per canister is:

Thrust ~ 2*Puncture area*Pressure
= 2*(0.0015 square inch) * 800 pounds/square inch
= 2 pound-force

The CO2 rocket force or thrust of 2 pound-force is much larger than the gravity force on a Pinewood derby car, at a 30 degree slope. And the CO2 rocket force keeps going until the gas runs out.

The main use of CO2 cartridges for pellet guns:

Pellet gun: Typical use for CO2 cartridge is a high pressure push for a pellet gun.

More typical chemical rockets, not CO2:

Fireworks

Estes toy rocket

Space rocket: the ultimate rocket with over 90% fuel.

Wide Nozzle: Strong, Quick Thrust

"Seeing is believing!"

Let's demonstrate some of the 'Arrow Principal' rules.

Top weight helps stability. We can place a heavy weight on the top of the rocket and see the straight-up stability of the launch. We show video frames of a slanted launch without top weight and a straight up launch with top weight.

In fact, with a moderate top weight, the peak rocket height does not change much because the bottle is now heavier and can plow through the air drag more easily. The bottle has a higher 'ballistic coefficient'. Of course, with too much top weight the thrust will not be much larger than the weight and the rocket will not go as high. We show measurements on the flight time as more weight is added to the top, and the flight time is flat, showing the trade between air drag, stability, and gravity.

Fits at the bottom help stability. We can make one of the fins a little slanted to cause instability, and show that the bottle tumbles without any top weight and does not tumble with top weight.

Sets of 4 quarters as weights added to top pouch of rocket

Other serious water rockets at higher pressures or higher flow rates:

Some hobbyists have made even higher pressure bottle rockets, at 500 psi. The PVC tube is used for the higher pressure, and the PVC tube is narrow to reduced air drag. A PVC tube has thicker plastic walls which handle the pressure. The thin walls of a bottle will not handle huge pressures.

Water rockets are even used as a joy ride as a tourist beach jetpack. People can fly on an unlimited water supply over a lake using a feed hose and water pump floating on the lake surface. On vacation at some lakeside beach, dare devils can now rent water jetpacks, in addition to sail boats and jet skis.

Jet pack: This water force is more than 200 lbs-force

Firemen need to handle the unintentional rocket of a fire hose, with huge water flow rates to put the fire out. Two firemen are needed to control the direction of the hose.

Fire hose

We have the 'Arrow Principal'. Sometimes you just need to verify it using other designs.

Water Launches and Straight Flights in Action
Wide Nozzle: Strong, Quick Thrust

"Now that's quick!"

Now this water blast is good exploding fun. There were many 'oohs' and 'aahs', and all around jumping, when the rockets were launched.
That water blast is clearly shooting out fast, and has drained the bottle within 6 feet in the air and a fraction of a second.   The flow rate from a large nozzle is fast.

Water Stops

Water gone right away

Water Starts

00:07

Quick blast with wide nozzle and rapid water exhaust rate.

Fill bottle with water and wide nozzle:
- 20-40% full

Pump:
- Up to 100 psi

Up forces:
- Push Water Out, an exhaust which creates thrust

Down forces:
- Gravity
- Air Drag

Instant Thrust:

For the wide nozzle design, the water gets expelled as quickly as possible. The water rocket is like a bullet, getting blasted out of the barrel by the gun powder extremely quickly. The faster the water leaves the rocket, the less energy is spent just raising the water with the weight of the rocket.

Of course, the trade, with real larger rockets, is that the payload must be able to take the jerk, or acceleration, without breaking apart. Human astronauts can only take up to 10 g's, whereas bullets can take much more than 1000s of g's and squish up against the rifled barrel.

Fun Facts about water thrust

All water coming out causes recoil forces, or thrust.

Water hits tree

Force back

Garden hose

Fire hose

Water jet pack

Faucet

See Appendix G for thrust estimates

Cool sprays of water and loud bangs happen as the pressure gets higher.
Don't overfill, or you'll just be taking away room from the air pressure.

Top Weight, Bent Fins, and Stability: Water Rocket Launch, Flips at 40psi

"Extra weight on top will allow a straighter flight for the water rocket too. And the longer thrust of the water can lift the softball higher."

The top weight softball also straightens out the flight even with bent fins, although the softball is so heavy that the water bottle rocket doesn't go as high.
The water exhaust can lift the softball higher than the air-only exhaust because the water exhaust lasts longer.

Flight is less stable with bottom weight:
- Water Rocket with no added top weight (no payload)
- Lower center of mass and lower pivot point, so bad

Flight is more stable with top weight:
- Water Rocket with added softball top weight
- Raise center of mass and raise pivot point, so good

Less top weight

Air flow

**Unstable:
Low center of mass: less stable**

Low center of mass, even farther down, where the water is!

Force back to alignment

Lever arm

Drag force because side of feathers are hitting the wind.

During tilt, the drag from side of feathers is too close to the center of mass, not enough to right the arrow, with less torque.

Frame 8
Frame 6
Frame 4

Curved path:
Without added top weight, there is less stability for flight direction.

Water Rocket, no top weight
With slightly bent fins, the un-weighted rocket flops around. There is no added weight at top of rocket to provide stability.

Frame 10
Frame 8
Frame 6
Frame 4

Softball on top

Water Rocket, with top weight
Even with the same slightly bent fins, the rocket goes straight, with added weight at top. A softball was the added weight at top of rocket to provide stability.

More top weight

Air flow

**Stable:
High center of mass: more stable**

High center of mass, closer to softball weight!

Lever arm

Force back to alignment

Drag force because side of feathers are hitting the wind.

During tilt, the drag from side of feathers is far enough from center of mass to right the arrow, with more torque.

Rocket needs the Center of Mass to be near top for straighter flight, which means added weight at top to counter act the water weight near the bottom.

High Pressure Uses of Water for Thrust and Energy

There are some amazing water rocket designs on the internet, beyond bottles. These designs instead use high pressure PVC pipes. These designs can apply a lot more pressure (500 psi), but they need higher electric pressure pumps! Bicycle pumps won't cut it.

Design of a very high pressure water bottle rocket:
- Long rocket to have more water fuel and energy
- Narrow tube to enable high pressures without breaking.
- Narrow tube for low air drag

"Some people get real pro about water rockets ...1700 feet altitude instead of 200 feet."

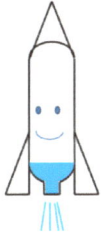

http://www.aircommandrockets.com/

Long launch tube for high tube pressure during initial launch

Height: 1750 feet

"Now that's some special tube, not just a plastic bottle. For really high pressure, the rocket launchers needed to ditch the bottle and bicycle pump, replaced by a high pressure cylinder and electric pump."

Very High Pressure Water Rocket: Here is a narrow rocket made from thick walled PVC pipe and pressurized with a high pressure hose at 500 psi.

Event 1:
Starting ground push from pressure in long tube

Event 2:
Rocket thrust from water exhaust

2-stage water rocket

Launch of High Pressure Water Rocket: 500 psi hose

http://www.aircommandrockets.com/flying_higher.htm

Steam Pressure in use

Steam Engine train: Hot steam pushes piston to power train wheels.

Steam engines

Aircraft Carrier catapult: Hot steam powered catapult, to allow heavy loads on the planes. The steam pushes a piston.

Steam catapult

Get some pent up steam and do a quick release.

Power Plants: Hot steam spins a turbine, to bring electricity to your house.

Steam for electricity

Much higher pressure, and a thin rocket, are of course great to get to higher altitudes.

"Use visual on height, or use flight time, and be impressed with how high the rocket goes."

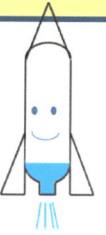

You can measure the height of the rocket in various ways, by comparing to a tree or building, or by measuring the time in the air. The height observation is a good way to see if you have a good rocket design.

This chapter asks the fundamental questions about how the bottle rocket works.
- What pushes the water exhaust out?
- What happens to the remaining air pressure after the water is already pushed out?
- Is air drag a big deal for bottle rockets?
- How much of your precious energy of pumping the bicycle pump actually goes into the height of the rocket or payload, as opposed to the launch height or raising of the water or the kinetic energy of the water?

To optimize something, it is helpful to know exactly how it works. First we have all the thrust events, and then we have the coasting or gliding through the air.

A bottle rocket actually goes through a few different thrust events, during that quick instant when the bottle is launching.
1. First there is the ground push from the tube. That is not a rocket thrust. It is simply the pressure from the tube pushing up against the bottle, supported by the ground.
2. Second there is the water exhaust. Water is pushed out at a rate that depends on the nozzle size and the pressure. More pressure means faster exhaust velocity and more thrust. Larger nozzle means more flow rate and more thrust.
3. Third there is the air exhaust of the remaining pressurized air. Maybe at first blush you'd think that air is much lighter than water and should produce less thrust, but, hey, these bottle rockets work with just air too. No, the air goes out at a much faster exhaust velocity than the water, so that creates the same thrust as the water did.

After all the pressure is gone and during projectile coasting, we have gravity and air drag acting on the bottle. Air drag is large. Because the bottle is so wide, because it is going so fast after the quick thrust, and because the bottle is now so light, the air drag is much larger than the gravity of the light empty bottle right after the thrust is done.

After a second or so, the bottle slows down enough due to air drag that the gravity is the dominant force on the bottle.

A bottle rocket can demonstrate that rockets are actually very inefficient, except of course for any other non-existent solution out there to get to space. For bottle rockets, the original energy is the muscle energy to pump the compressed air. Some of that energy goes to heat the gas, and the rest goes to the pressure. During launch, the rocket needs to lift the water and that wastes energy. With exhaust, the exhaust itself has energy of motion, and that wastes energy. Rockets have a hard life.

Air
Stops

Water
Stops

Water
Starts
Tube
pressure

Water profile at 100 psi launch

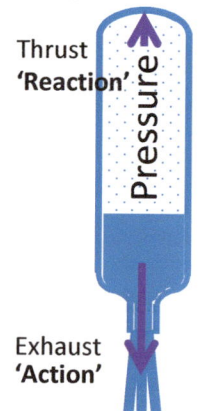

Thrust
'Reaction'

Pressure

Exhaust
'Action'

'Reaction' from pushing out water and gas

Here are two ways to estimate the maximum height of your rocket. One's a comparison to a tree or building, and the other is to just measure the flight time.
More height is the result of more thrust or longer thrust, and a well designed rocket with straight flight and less drag.

Height and Flight Time

Of course you want to know how high your rocket goes. Deep inside your shameless and competitive instinct, you're either bragging about the ugliest rocket, worst crash rocket, biggest rocket, or fastest rocket, but you're also shooting for the highest rocket.
If you're standing right under the rocket, it is really hard to tell the height. You'll probably need to measure the hang time. Or, if someone else can stand far away, that person can get the height by comparing the rocket apex to a tree.

"Use visual on height, or use flight time, and be impressed with how high the rocket goes."

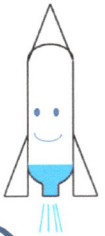

Measure height directly, compared to a tree :

The most direct way to measure the height of the flight is to stand back and compare the height of the rocket to the known height of a tree.

2 tree units, 14 person units, or 70 feet

Use tree as reference for height.
Back up a little farther!

1 tree unit, 7 person units, or 35 feet

I'm 5 feet tall.

Stand far away and use a tree or building as a reference.

Measure time of flight to estimate height, using a watch or cell phone :

Here is a way to estimate the height of the flight of the rocket, using a simple equation. Just measure the time of flight, and now you have an estimate of the height, using Littlewood's rule.

- Highest launch of the day took 7 seconds, at a little over 100 psi.

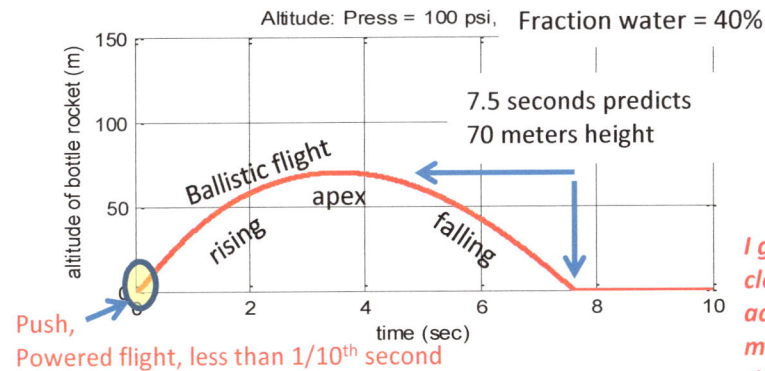

Altitude: Press = 100 psi, Fraction water = 40%

altitude of bottle rocket (m)

Ballistic flight
apex
rising   falling

7.5 seconds predicts 70 meters height

Push,
Powered flight, less than 1/10th second

I get to be close to the action, measure flight time, and use my math skills!

Time (second)	Height (meter)
5 sec	31 m (100ft)
7 sec	61 m (197ft)
9 sec	80 m (326ft)

Littleton's rule:

$$height_{apex} = (g/8)(time_{flight})^2$$

where gravity g = 9.8 m/s^2

- Time of flight rule can be used for bottle rockets, toy chemical rockets, arrows, all with no parachute.
- Time of flight rule assumes the rocket is mostly going up.

Ballistic Flight: Like a bullet with no power after launch, the bottle just uses the speed and momentum already given by fuel, and coasts against gravity and air drag.

Ballistic examples:
- Bullet, after leaving the barrel
- Rocket, after fuel gone
- Satellites orbiting earth, constantly falling

Not Ballistic examples:
- Rocket propelled grenade
- Airplane
- Rocket getting into orbit.

Here are two ways to estimate the maximum height of your rocket. One's a comparison to a tree or building, and the other is to just measure the flight time.

Wide Nozzle: Strong Thrust, Quick Launch

Video and pictures of the water exhaust provide evidence of different exhaust steps, and directly tells the height of the water discharge. The same video snap shots are also possible with toy chemical rockets.

"There are 4 different timelines to the water flight."

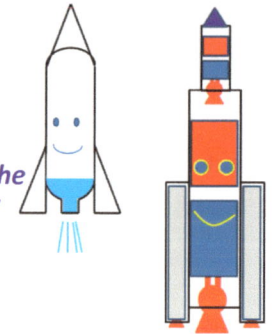

Multi-process analysis of the water rocket launch, one event after another

"Well, I have 3 stages and many timelines for my chemical burn, dropping hardware as I go."

Newton's third law:
- Action and Reaction: For every force, there is an equal and opposite force.

Water gets pushed down, and that pushes the rocket up.
The air pressure in the rocket pushes on the water. Because there is an equal and opposite reaction, the water is pushing up on the air in the rocket when the air pushes down on the water. That accelerates the rocket.

This water snapshot is just remarkable.
On the tube at bottom, there is the water scattering off the launch tubes. With water exhaust at middle, there is the jet stream of water. With air exhaust at top, there is the mist of water that comes out just as all the water is gone from the bottle. The high pressure air is still escaping from the bottle, which disrupts the water and forms the mist on top. The three events, the tube push, the water exhaust, the air exhaust, all add thrust to the rocket.

The launch is fast.
For a wide nozzle like this launch, the water is all ejected within 0.1 seconds due to high flow. Water is ejected faster at higher pressures.

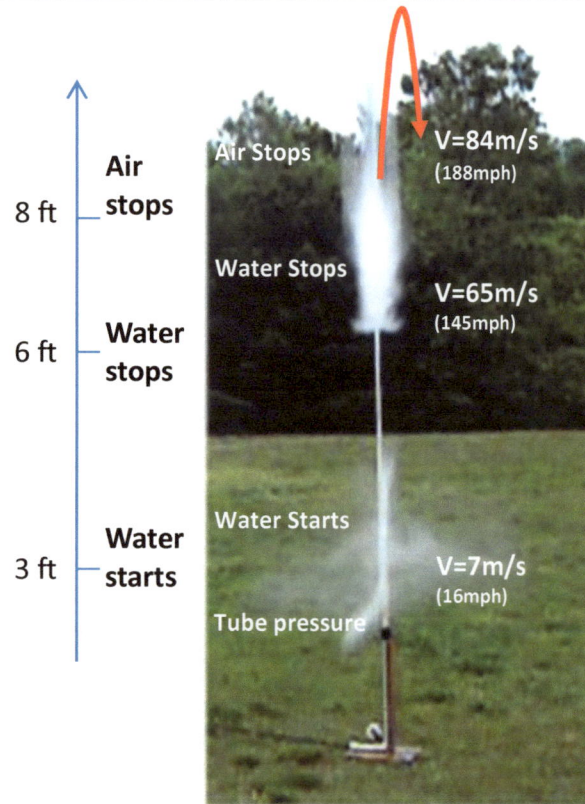

Event 4: Coasting with free un-powered ballistic flight

Event 3: Air ejection from bottle, still have propulsion

Event 2: Water ejection from bottle, for main propulsion

Event 1: Air push from ground launch tube

Experiment: Take high speed video of launch and see if can identify all events (tube pressure / water exhaust / air exhaust)?

Timeline and camera frame rate:
The water remained visible for a long time in the video snap shot. The video rate for this standard digital camera is 25 frames per second, so this water at least stayed in this action packed shape for at least 40 milliseconds.

Photo labels
- Air stops — V=84m/s (188mph)
- Water Stops — V=65m/s (145mph)
- Water Starts — V=7m/s (16mph)
- Tube pressure
- 8 ft — Air stops
- 6 ft — Water stops
- 3 ft — Water starts

Water profile at 100 psi launch

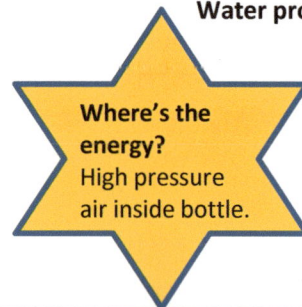

Where's the energy? High pressure air inside bottle.

For every force, there is an equal and opposite force. This is the principle of action and reaction, or Newton's 3rd law.
→ Water and air go down and out. The rocket goes up.
→ For guns, the bullet goes out. The gun gets recoil.

$Mass_{pellet}$ v_{pellet}

$Force_{back}$ $Force_{pellet}$

← The gun and shooter and floor get pushed backward
← Rocket thrust

→ Pellet gets pushed forward
→ Exhaust gases (water, air)

Even though the exhaust water is cold, the exhaust is still like a chemical rocket. For all rockets, for each action, there is a reaction.

Stages of Water Launch: Quick Launch as Water Blasts Out

Here are the beginning thrust and coast stages of the water rocket launch. Using a PVC tube launcher, the water thrust is the event 2 below after the rocket rises above the tube, and the free space coasting with air drag is event 4.

Event 4, un-powered flight (ballistic):
Gravity and air drag are in charge, with no more exhaust and thrust. The rocket is already pushed as fast as it is going to get, with no more thrust, just inertia.

- Here is the free un-powered flight stage, or ballistic stage. The rocket is not going to get any faster, because all the propellant is gone (water and high pressure air), so the rocket will keep coasting up until gravity and air drag cause it to come back down to earth.

Event 3, Air exhaust and thrust (F=2*P*A):
Then, from 5 feet to 10 feet, the high pressure air is pushed out of the bottle rocket, giving extra thrust.

- Water is gone and water thrust is gone, but there is still air pressure inside the bottle. Now we basically have an air rocket, at about half the initial pressure of the pumping. The pressure was reduced because the air in the bottle expanded pushing the water out.

Event 2, Water exhaust and thrust (F=2*P*A):
In the first 5 feet, for a wide nozzle typical plastic bottle, all the water is pushed out of the bottle rocket from air pressure quickly, for the main thrust.

- Here is the main propulsion, using the ejected water. The water may not be ejected at super high exhaust velocity, but water is heavy, and the change in momentum really gets the water rocket moving.

Event 1, Air pressure push from launch tube (F=P*A)

- The tube inside the water rocket is above the water level, and the air pressure can push against the rocket. This is only for about 6 inches of rise of the bottle along the tube.

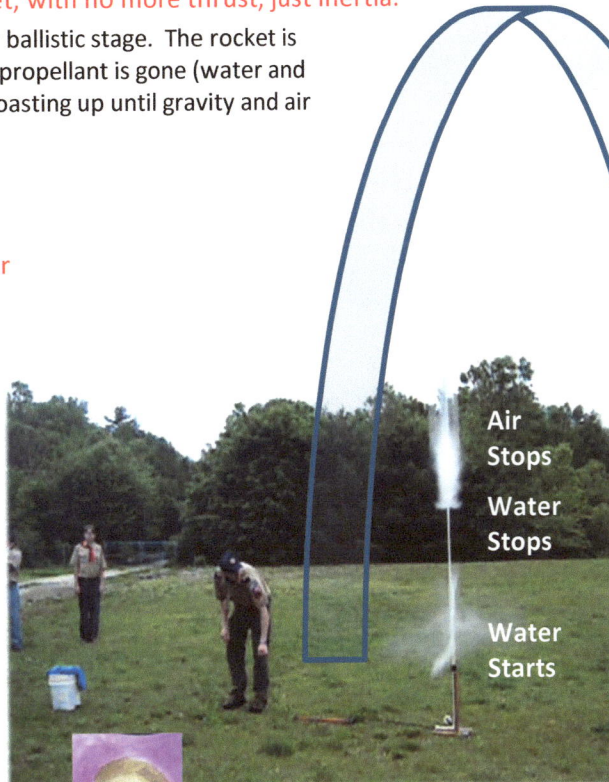

"I've got some push, but gravity keeps me down."

"Well, I've got 3 stages, chemical fuel, and high mass ratio, so I get to orbit."

Event 4: Free un-powered flight (ballistic flight)

Event 3: Air ejection, still have propulsion

Event 2: Water ejection, main propulsion

Event 1: Air push from launch tube

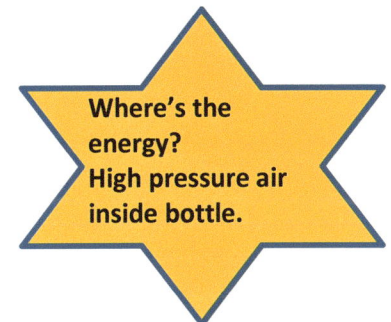

Air Stops

Water Stops

Water Starts

Launching at a Scouting event

High pressure compressed air pushes water out as exhaust which pushes rocket

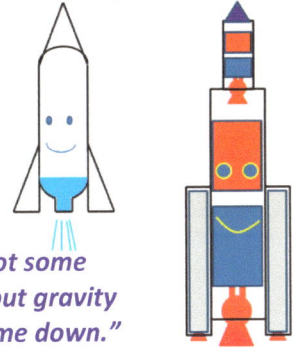

Where's the energy? High pressure air inside bottle.

Most of the flight is coasting. The first three thrust events only take about 0.1 seconds for a wide nozzle. The exhaust thrust is 2*Pressure*Area, and it doesn't matter if the exhaust is air or water. But water exhaust lasts longer.

Life of Water Rocket: 3 Events of Thrust

Here are three events of water bottle thrust – ground push from air tube pressure, thrust from water exhaust, and thrust from air exhaust. The water exhaust is the dominant one to get to higher speeds because it lasts longer, but the air tube pressure and air exhaust still contribute, just like an air bottle rocket.

"Sometimes you just need to break the problem down into smaller steps."

The three events that happen during thrust, one after another

Event 1:
Tube push from pressure in tube

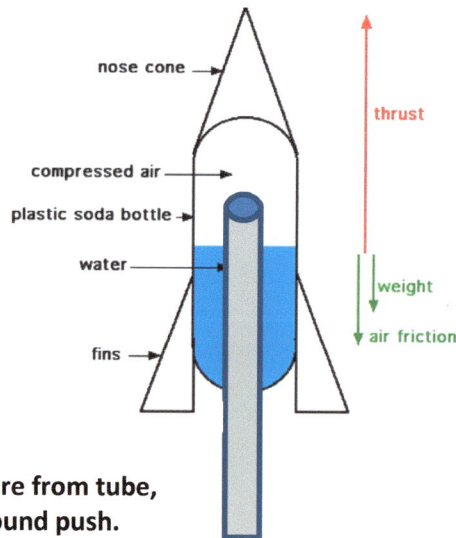

nose cone

thrust

compressed air

plastic soda bottle

water

weight

air friction

fins

Pressure from tube, for ground push. During this first even, no water is expelled yet.
The rocket is still attached to the ground and can push off.

Force = Pressure*Area

Event 2:
Expel water for thrust

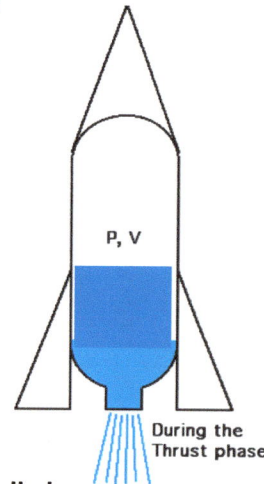

P, V

During the Thrust phase

Water expelled for rocket thrust:
Exhaust water has the longest duration so it causes the main thrust.
The bottle is now a rocket.

Force = 2*Pressure*Area

Event 3:
Expel remaining air for thrust

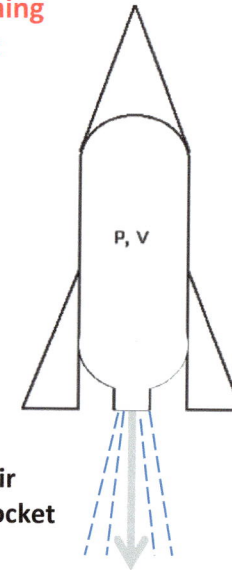

P, V

Compressed air expelled for rocket thrust, at remaining pressure.
The bottle is still a rocket until the air exhaust is gone.

Force = 2*Pressure*Area

We get about 30% more velocity by accounting for all the thrust events — air tube, water, last air exhaust — not just the water exhaust.

Pressure, Water Exhaust, and Air Exhaust

The internal air pressure keeps dropping as the air expands and pushes out the water and then its own air.

As the water is pushed out, the air volume expands and the pressure drops. So the water exhaust velocity does drop a little during the water thrust time.

When the water is all gone, the air itself is forced out of the nozzle. The air speed is much faster than the water speed because the air is less dense. There is still the same amount of thrust, up until all the air is gone.

"All that air pressure is doing work."

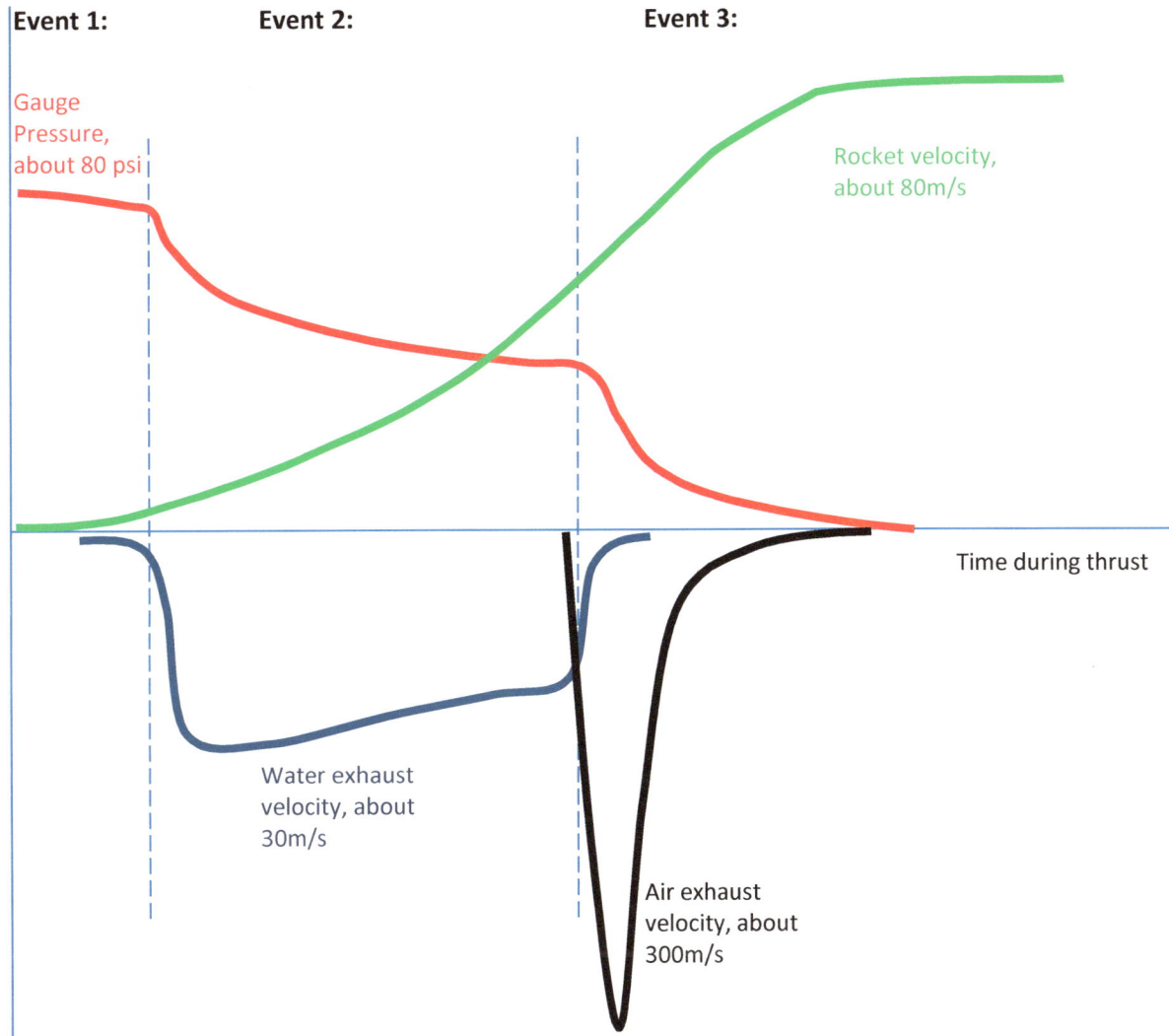

Event 1:

Event 2:

Event 3:

Gauge Pressure, about 80 psi

Rocket velocity, about 80m/s

Time during thrust

Water exhaust velocity, about 30m/s

Air exhaust velocity, about 300m/s

How air pressure drops during the quick thrust to launch the rocket to its final speed.

There are the 3 events of the thrust. First the tube push, then the water exhaust, and last the air exhaust.

Air Drag on Water Rocket is Larger than Gravity

Here are all the thrusts, gravity, and drag forces on the rocket, with directions of the forces, as the rocket launches and coasts through the air. The thrust needs to be larger than gravity to even get off the ground. For a wide nozzle the thrust can be 50 times larger than the down force due to gravity. Drag is about 2 to 5 times larger than the force of gravity, at the fastest speed just after launch when thrust has just finished, with a typical bottle with a wide diameter.

"The water thrust is only the small beginning part of the trajectory or arc. Mostly the rocket is coasting like a baseball. What goes up, must come down, unless it achieves orbital velocity."

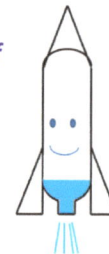

Air Drag

Time ~3 seconds

Gravity

Air Drag

Gravity

Huge initial push from water exhaust

Air Drag

Gravity

Coasting:
No more exhaust and thrust, just momentum, like an arrow.

Air Drag

Gravity
(0.5 lb-force)

Air Drag

Gravity

Launch:
Exhaust and thrust time ~0.1 seconds for wide nozzle

(50 g's)

Gravity
(3 lb-force)

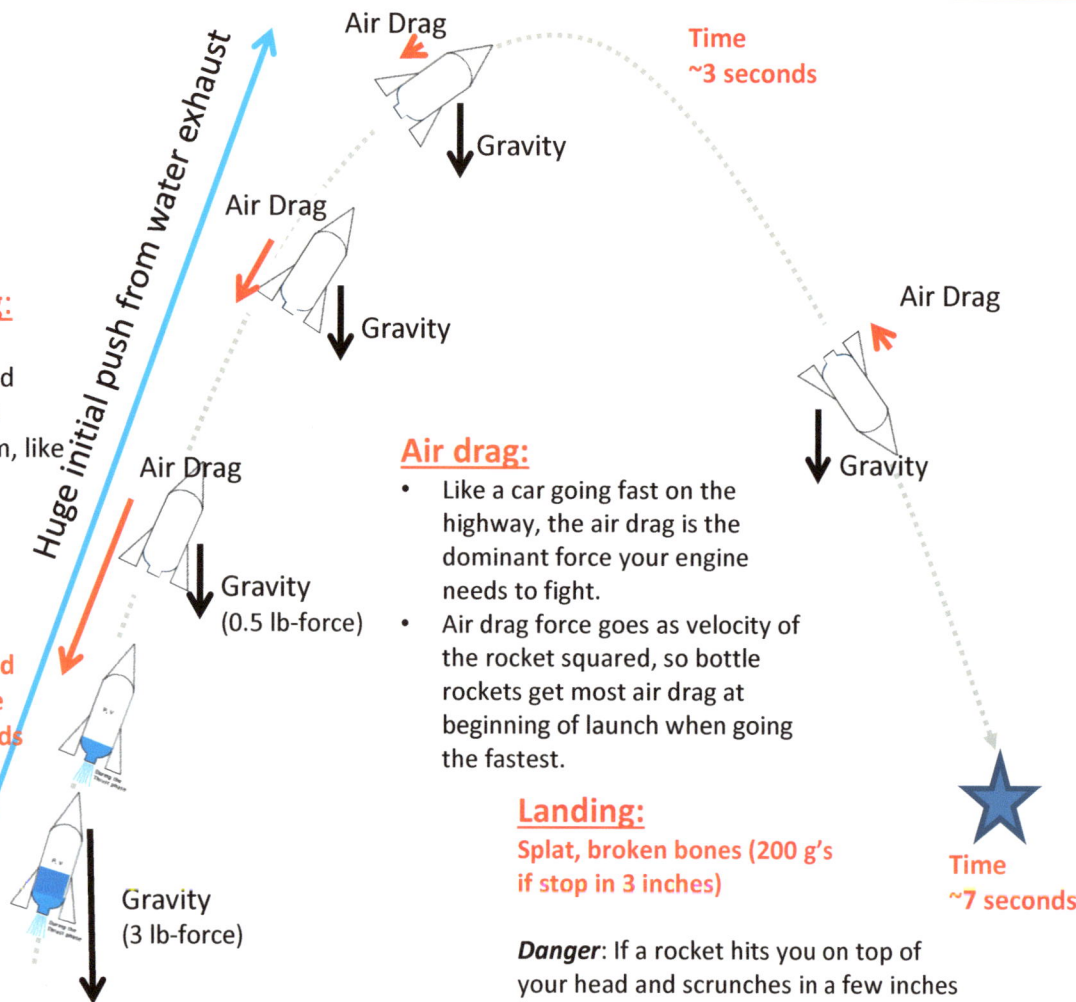

Umph! What a kick in the gut. Hard to track launch with the eye. Need a slow motion camera.

Air drag:
- Like a car going fast on the highway, the air drag is the dominant force your engine needs to fight.
- Air drag force goes as velocity of the rocket squared, so bottle rockets get most air drag at beginning of launch when going the fastest.

Landing:
Splat, broken bones (200 g's if stop in 3 inches)

Time ~7 seconds

Danger: If a rocket hits you on top of your head and scrunches in a few inches (a crumple zone), that's equivalent to 100 lb weight. So clear the launch area, and aim the rocket away from people.
- From Newton's F=ma, for a ½ lb rocket.

Question: How efficient is this water bottle rocket?

Answer: Like chemical rockets, most of the energy goes into the fuel (water) pushed out, and in raising the fuel (water) before the fuel is pushed out. Very little of the energy goes into the 'dry rocket', the plastic bottle payload.

This low efficiency, or huge amount of fuel and small payload, is why launches cost $1000s per pound of payload to launch anything into space.

Other reasons for high space rocket costs are expensive one-shot parts, quality control, and constant maintenance.

Hand out car window with air drag
Feel the drag on your hand out the window of a car. The drag force on your hand is equal to the force it takes you to hold your hand against the wind.

Bottle rockets are going up to **100 mph** right after the water is expelled, so the air drag force is very strong.

Drag force is more than **2 lbs-force** on an empty rocket of less than 0.5 lbs at 60 m/s (>120mph).

Most the flight time of the bottle rocket, after a quick water blast, is coasting, with gravity and air drag slowing the rocket down.

Air Drag, and Sports

Air drag on game balls impacts a lot of sports.
Round balls, perfect spheres like baseballs and basketballs, have lots of air drag for their size. Footballs, with an oval shape, have less air drag for their size.

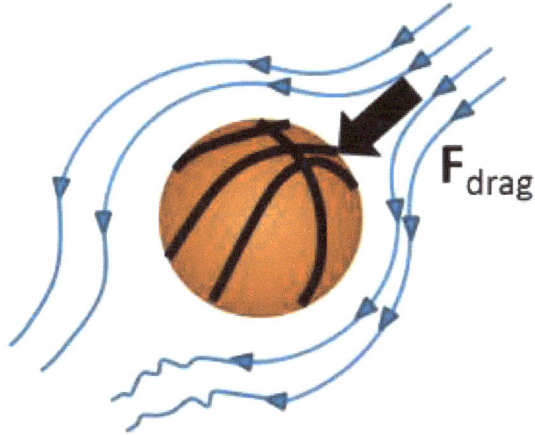

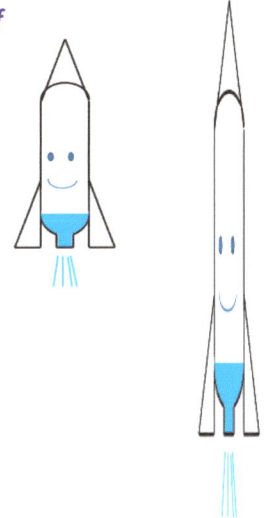

F_{drag}

F_d

F_g

Basketball shots need to account for air drag. A game in the mile high city in Colorado in the mountains has less air drag where the air is thinner, compared to a game in Florida near ocean level. Players automatically adjust their shot.

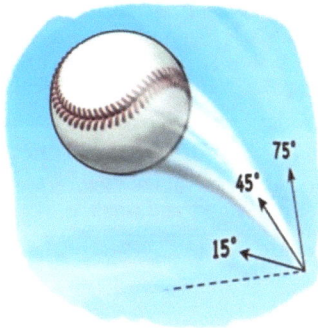

An objects falling will reach a terminal velocity, where the force of gravity is equal to the force of air drag. There is no more increase in velocity down after that.

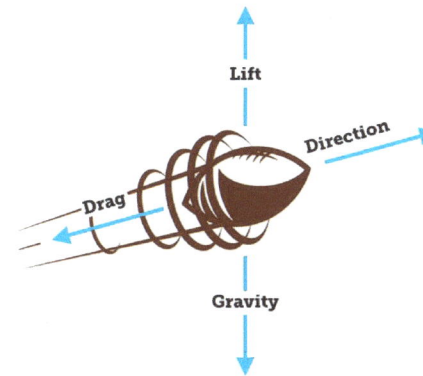

75°
45°
15°

Lift
Direction
Drag
Gravity

A baseball will go farthest when hit at 45 degrees in a vacuum. With air drag, the speed is slowed down right away and it is better to hit at a lower angle to get the ball farther away right away. For example, a whiffle ball with huge air drag should just be thrown straight.

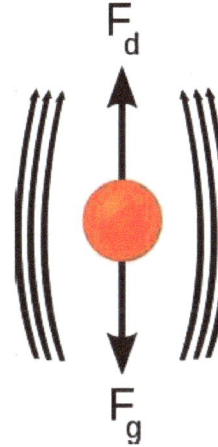

A football has a more aerodynamic shape, with less air drag for the size of the ball. But now you need to throw a spiral.
A football, with its oval shape, will keep its forward speed and energy for longer.

Balls actually have a lot a air drag for their size. A football shape is better for the same size, We use aerodynamic shapes to reduce drag coefficient for the same area object.

This chapter shows videos of many rocket launches – water bottle, toy chemical, and space chemical – and figures out the acceleration and thrust.

For a rocket to rise up against gravity, it needs a thrust greater than its weight. During a launch, an acceleration of 2 g's requires a thrust of 3 times the weight because gravity is still pulling back. Out in space, without Earth's gravity, a thrust of 3 g's will cause an acceleration of 3 g's.

The motion of the rocket can be easily seen. Just watch it launch, or even record the launch on video.

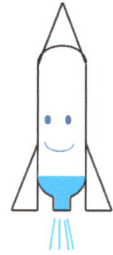

What determines the acceleration of the launch?

Toy rockets tend to be small and the thrust tends to be huge compared to the weight of the rocket. The launch is much more dramatic that way. However, due to the little amount of fuel, or due to the low energy content of the fuel like compressed air, the fuel runs out very quickly and the toy rocket falls back down.

Videos of the launches show the change in altitude of the rocket during the launch. This is a direct measurement of the thrust.

First the thrust needs to be larger than the weight of the fully fueled rocket, or larger than gravity. If the thrust equals the weight of the rocket, that is equivalent to the rocket just resting on the ground, with the ground replaced by the thrust.

Let's say we know an average thrust, and we know an average mass of the rocket during launch. Then we can plot the acceleration and the velocity versus time and the height versus time during the thrust. These are basic equations of motion, and these plots are shown at the end of this chapter.

A large nozzle will cause a larger thrust, which means more g's. A bottle rocket can accelerate at up to 50 g's with the high bottle pressures of 100 psi and a 25% fill of water.

Real space rockets have g's of only 3 or less because people are on board, and because the rocket needs to light with less structural support.

Impulse of the fuel:

What is the relationship between the total momentum change of the fuel, or the 'impulse', and the final speed of the rocket?

Ignoring gravity, the final speed of the rocket does not depend on the instantaneous thrust. The final speed is the product of the instantaneous thrust and the time duration of the thrust, which happens to be the 'impulse'. So the rocket could have a slow burn rate and still get to the same final speed, because the time duration is longer.

When it comes to real space rocket design, we do care a lot about instantaneous thrust for the first stage. The first stage is all about getting off the ground with the full weight of the rocket. So the thrust needs to be more than 2 times the weight.

What is the impulse of the water in the bottle? Impulse is exhaust velocity and mass, so just multiply the water mass by the exhaust speed, which depends on the air pressure.

We always care about the impulse of the fuel. That is another parameter that means high energy density, or high specific impulse.

For space rockets or water bottle rockets, we need the rocket equation and the rocket equation says that the final velocity is mostly dependent on the exhaust velocity, not the flow rate or even the mass ratio.

"Rockets accelerating at 1 to 3 g's is a good sweet spot for human astronauts, without feeling like a bug splat on a windshield."

Videos will show height versus time, or acceleration.

Smaller nozzle:
- **Small flow and small thrust**

Medium nozzle:
- **Medium flow and medium thrust**

Larger nozzle:
- **Large flow and large thrust**

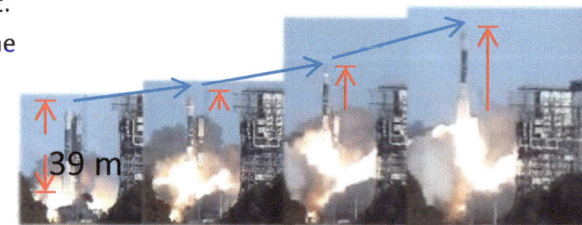

39 m

Snapshots of space chemical rocket launch.

Toy rockets can have huge g's, above 10 g's, because no astronaut is on board and the cardboard or plastic is thick. Space rockets need to keep to a few g's to be safe for people and not break the rocket apart.

Thrust and Acceleration, Videos Reveal All

Videos of many rocket launches – water bottle, toy chemical, and space chemical – provide a way to measure the acceleration and thrust. It is easy getting the rocket height versus time. Just take a video and look at the rocket height from frame to frame.

The rocket needs exhaust and thrust to go anywhere. With a launch video or by eye, you can watch how fast the rocket rises and use Newton's 2nd and 3rd laws to get the thrust. For example, use a stopwatch and some known comparison heights, like a tree, the launch tower, or the rocket itself.

Thrust, or force from exhaust, must be larger than the weight of the rocket in order for the rocket to rise vertically off the ground. A lot of failed launches still have some thrust, but the thrust is less than the weight of the rocket.

Out in space, where the rocket is travelling so fast that it is in orbit, the thrust can be anything, even weaker than the weight of the rocket on Earth. For example, burns for orbital corrections, or 'station keeping', use a thrust less than the weight of the rocket, but for long burn times.

For thrust of water bottle toy rockets, depending on the size of the hole at the bottleneck, the thrust can be anywhere from 2 times the weight of the filled rocket with narrow bottlenecks and low flow rate, to 50 times the weight with wide open bottlenecks and fast flow rate.

For thrust of chemical space rockets, the thrust is typically 2 to 4 times the weight of the rocket. The nozzle expands to direct exhaust downward, using a converging/diverging nozzle design.

Water exhaust:
- About 300 pounds thrust, equal to the weight of the pilot and pack

Static Test of Thrust for Estes toy rocket engine:
- About 1 pound thrust, pushing up a 0.1 pound rocket

Thrust and acceleration are directly related, using Newton's 2nd law:

$$Force = Thrust = Mass * Acceleration$$

So when you record the acceleration of the rocket using a camera, you are also recording the force or thrust.

The thrust for a rocket can be expressed as, using Newton's 3rd law:

$$F_{rocket} = (v_{exhaust})(mass\ flow\ rate)$$

or

$$F_{rocket} = v_{exhaust}^2 Area_{nozzle} \cdot density_{exhaust}$$

For exhaust velocity of toy bottle rockets, a higher pump pressure increases the exhaust velocity, for more thrust. Also, a larger diameter bottleneck or nozzle allows more water flow, so more thrust.

For exhaust velocity of chemical rockets, the heat of the chemical reaction keeps a high gas pressure, which is then directed out the nozzle.

The different rocket stages have different nozzle designs. Nozzles that burn in outer space have a larger flare than nozzles that burn in low Earth atmosphere. This nozzle flare is based on sideways exhaust pressure equal to the outside air pressure, so that the exhaust can go straight down for most thrust.

Water bottle: O-ring launcher at 25 psi.

Water bottle: Homemade PVC launcher at 50 psi.

**3...2...1...blast off
(T minus 3, minus 2, minus 1)**

Static Test of Thrust for space rocket booster:
- 3.5 million pounds thrust, need a few of these to lift a 6 million pound rocket

Exhaust and Newton's 3rd law of acceleration go hand in hand

Launch and Acceleration: Video of Water Bottle Rocket with Wide (full 7/8th inch opening) and Narrow (1/10th inch) Nozzle

Water bottle rockets demonstrate the same thrust ideas as huge space rockets. Relative to their weight, water bottles can have huge thrust, or just gradual thrust, depending on the pressure, the size of the nozzle and the flow of water. Wider nozzles have more water flow and more thrust.

Wide nozzle: 50 g's (thrust 51 times the weight of the rocket) … like a car crash

Full open nozzle and more flow and thrust…fast!

"This wide nozzle bottle rocket has a mean kick-start."

Frame 1 — height=0 T=0

Frame 2 — height=0.4m T=0.04 sec

Frame 3 — height=2m T=0.08 sec

This would hurt if the rocket hit you! An elite boxer's punch is around 1000 lbs-force, moving your head back at 50 g's.

With a wide nozzle and huge water exhaust flow, the water bottle rocket has 50 times more acceleration than gravity!

These water rocket accelerations, of 50 to 100 g's, are like a car crash, and would squish and kill people, astronauts, if they were in the rocket.

Narrow nozzle: 1.1 g's (thrust 2.1 times the weight of the rocket) … like space rocket

These water rocket accelerations, of 1 to 3 g's, are comfortable for people if they are lying horizontal in the rocket.

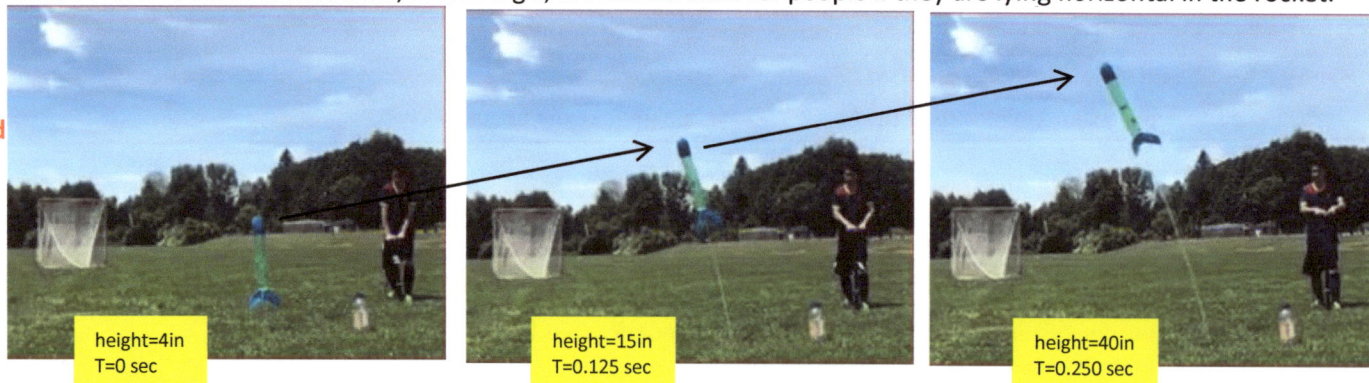

Narrow nozzle and less flow and thrust…typical gradual 1g space rocket or NASA accelerations

height=4in T=0 sec

height=15in T=0.125 sec

height=40in T=0.250 sec

Using cell phone video on slow-motion mode: 240 frames per second (0.005 seconds per frame)

With a narrow nozzle and small flow, the water bottle rocket has lower thrust and acceleration closer to gravity.

The smaller flow from the smaller diameter of the nozzle causes less thrust, but the rocket gets to the same height over time because the smaller flow water exhaust lasts over a longer time and because there is less air drag.

Launch and Acceleration: Videos of Higher Energy Chemical Rockets

The acceleration of rockets at launch is given by the ratio of thrust to weight.
Chemical rockets also can have fast launches for toys or missiles from 10 to 50 g's, or can have gradual launches for space rockets at about 2 to 3 g's. Fast launches are safe when no people are on board. Gradual acceleration does not harm astronauts and allows thinner structures and less hardware weight.

Quick push from toy engine: 20 g's for toy rocket! (thrust 21 times the weight of the rocket) … not like space rockets

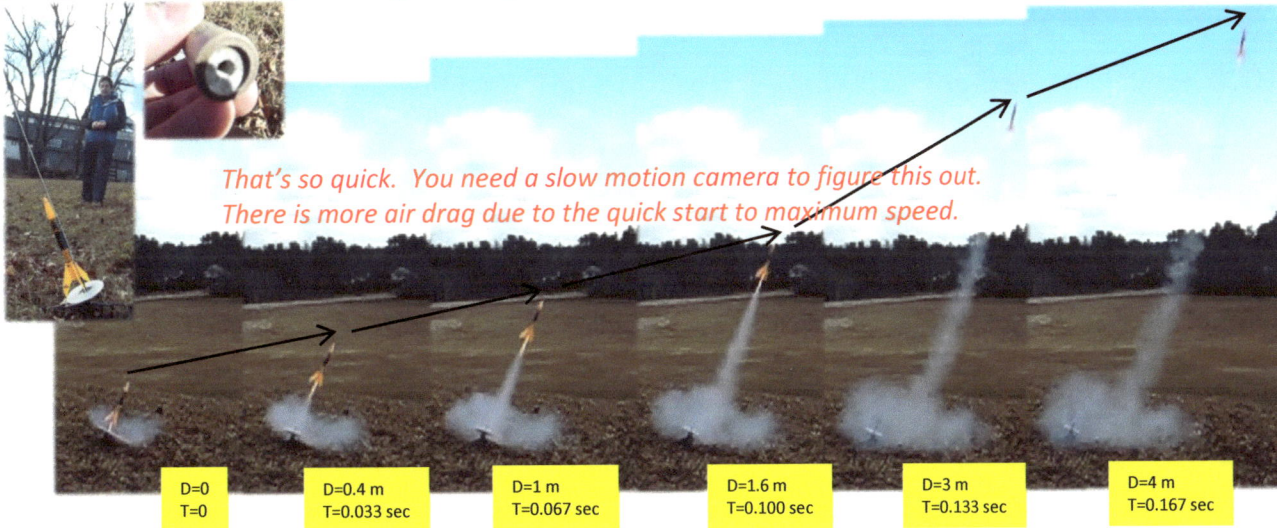

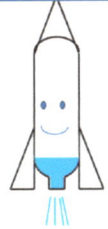

That's so quick. You need a slow motion camera to figure this out.
There is more air drag due to the quick start to maximum speed.

"Boom! And the rocket is gone."

These accelerations of toy rockets, of >20 g's, are like a car crash, and would harm or kill people if they were in the rocket.

| D=0 T=0 | D=0.4 m T=0.033 sec | D=1 m T=0.067 sec | D=1.6 m T=0.100 sec | D=3 m T=0.133 sec | D=4 m T=0.167 sec |

30 frames per second
(0.033 seconds per frame)

Gradual pickup for space: 1 g's for space rocket (thrust 2 times the weight of the rocket) … this is exactly a space rocket

A human-friendly, tame acceleration of about 1 g.
There is less air drag by gradually picking up speed.

Delta 2 rocket launch of NASA MITEX
spacecraft from Cape Canaveral, FL
Rocket length: 39 meter length (128 feet)

39 m

The accelerations of 1 to 3 g's are tolerable for living astronauts. If thrusts and accelerations were larger, then people would feel a huge force on them and have trouble breathing.

| Height = 0 meters T = 0 sec | Height = 8 meters T = 1 sec | Height = 22 meters T = 2 sec | Height = 40 meters T = 3 sec |

Chemical rockets have a lot of energy. The space rocket is designed for 8 minutes of steady thrust into orbit and outer space, with 2 or 3 stages. The toy rocket is designed for fun, less than 0.2 seconds of thrust, and below 1000 feet.

Water Has Mass, But Does Air Have Mass, to Shoot Out?

An air only rocket would not work unless air had mass, from the gas molecules. Air drag would not happen unless air had mass. Yes, an air-only bottle rocket has mass in the compressed air. That compressed air is where the energy and exhaust are, so compressed air better have mass. We're only talking 0.022 lb of compressed air at 100 psi in the bottle, but that is the energy and the exhaust. For an air-only bottle rocket, because the plastic bottle weighs only 0.27 lb, the mass ratio is only 1.1.

Yes, a pressured scuba tank will feel heavier after it is pumped up (6 lbs heavier just from compressed air), but not heavier than the thick metal walls that hold the 3000 psi pressure in. Scuba tank walls are intentionally heavy to be neutrally buoyant in water. Space rockets are also heavy, but mostly from the fuel.

Below shows the mass per liter for 1000 psi compressed gas, 3000 psi compressed gas, and liquid cryogenic fuel.

Bottle Rockets:

Air mass: A 2 liter bottle at 100 psi will have 10 grams (0.022 lb) of compressed air.

Water mass: At 40% full, have 0.8 kg (2 lb)

Pressure from bicycle pump:

- Compressed air to 100 psi.
- Bottles need to keep half the bottle volume to contain air for steady pressure.

Scuba Tanks:

Air mass: A typical scuba tank will have 3000 grams (6.4 lb) of compressed air, in a tank holding 100 cubic feet of air (2000 liters).

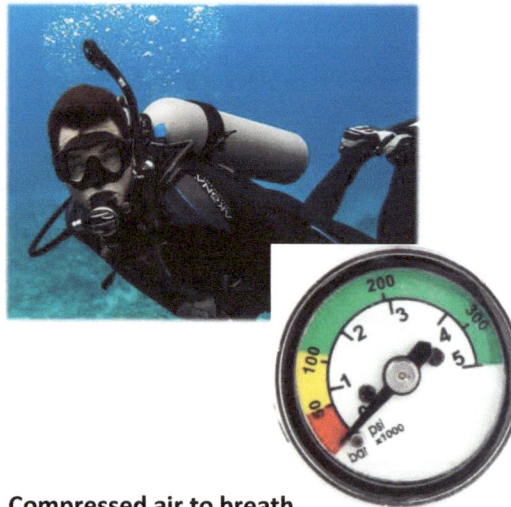

Compressed air to breath underwater:

- Compressed air up to 4000 psi in a scuba tank.
- Divers use more pressure to allow more air and a longer dive time.

Space Rockets:

Liquid mass: At launch a space rocket will have >1Million lb (>0.4Million kg) of liquid fuel and oxidizer at cryogenic temperatures. Liquid oxygen comes from air.

Liquid oxygen at very cold temperature

Liquid oxygen storage tank:

- Liquid oxygen is pumped into rockets on launch pad just before launch. The liquid oxygen is -218 degree Celsius (55 degrees Kelvin).
- This tank is over 100 feet tall.

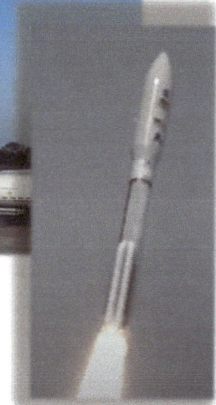

Atlas V rocket :
Liquid kerosene fuel / oxygen first stage using Russian engine, and liquid hydrogen fuel / oxygen second stage

Air has mass. Even after the water exhaust is gone, there is still air pushing out the nozzle, until the bottle pressure equals the outside pressure.

Thrust, Burn Time, and Final Rocket Velocity

To get to the same speed, do you want to accelerate fast for a short time, or accelerate gradually for a long time? Out in space it does not matter, a quick or a longer duration. The final velocity is the same, whatever your choice for acceleration profile, for the same amount of fuel.

Final rocket velocity does not depend on the thrust or the exhaust time (burn time) exactly, but instead the product of the two.

Same impulse or amount of fuel

Thrust and time

Thrust versus time curves, with rocket velocity change, for long thrust and for short thrust.

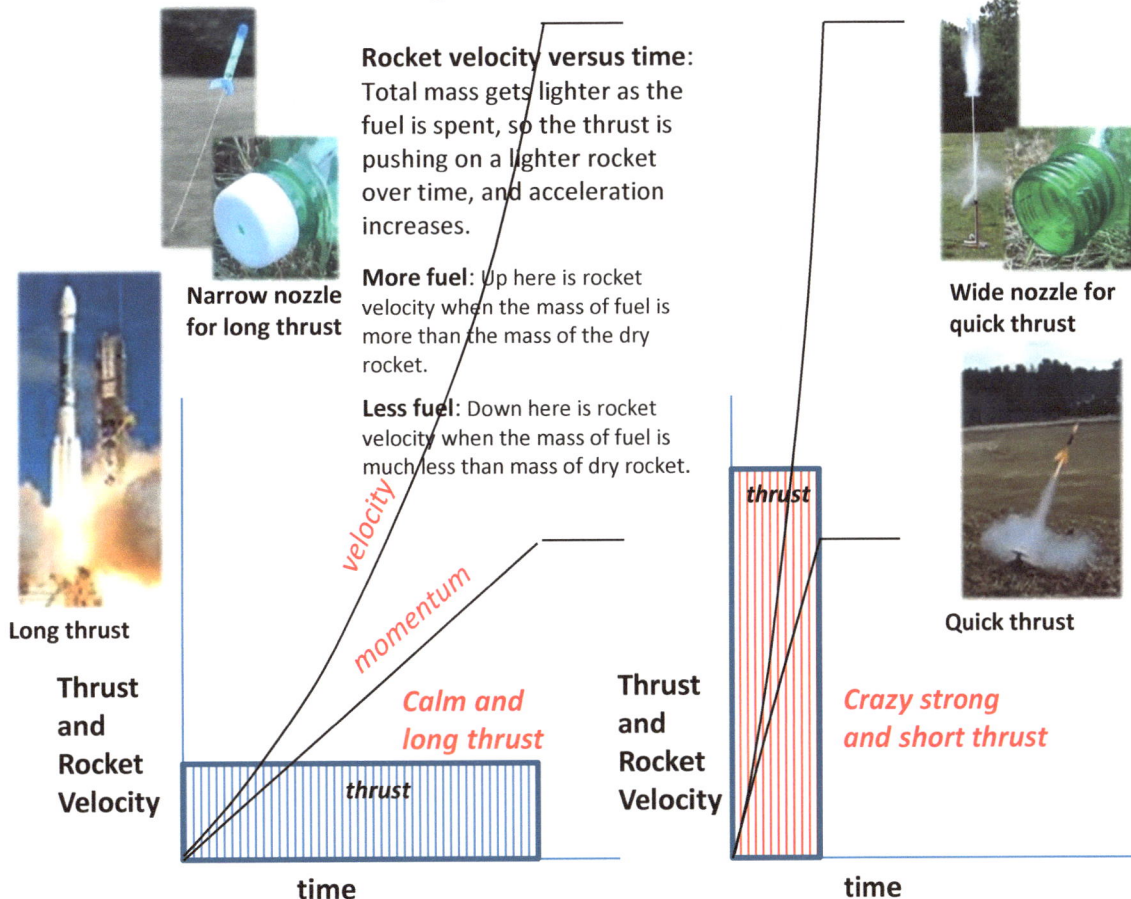

Rocket velocity versus time: Total mass gets lighter as the fuel is spent, so the thrust is pushing on a lighter rocket over time, and acceleration increases.

More fuel: Up here is rocket velocity when the mass of fuel is more than the mass of the dry rocket.

Less fuel: Down here is rocket velocity when the mass of fuel is much less than mass of dry rocket.

Narrow nozzle for long thrust

Wide nozzle for quick thrust

Final rocket velocity does not necessarily get faster when the thrust is larger but short duration. The final velocity is thrust times the burn time, not just thrust.

Quick thrust

Long thrust

Thrust and Rocket Velocity

velocity

momentum

Calm and long thrust

thrust

time

Thrust and Rocket Velocity

thrust

Crazy strong and short thrust

time

Area under the curve is the total change in momentum of the rocket, which gives the final rocket velocity.

*momentum change = thrust * time = mass * final velocity*

Force and time examples:
Think about a race between a super-charged race car and a small under-powered car. Both accelerate to 60 mph, and stop going faster, or else the car is speeding. The race car has the power and torque to go from 0 to 60 in 3 seconds. The under-powered car has weaker torque and takes 10 seconds. But both get to the same speed, so both cars get the same momentum change, or impulse.

Think about the opposite, the reverse deceleration to stop a car, going from 60 to 0 mph. You can decelerate gradually over 100 meters, or dangerously and suddenly over 1 meter. The 100 meters is a gradual stop and lower force. The 1 meter sudden stop is a crash, where the force is huge. Both have the same velocity and momentum change. One hurts!

Same energy for both fast burn and slow burn, and same final velocity.

Thrust and Final Velocity, at Different Burn Rates

The thrust magnitude is part of the rocket design. There is only so much fuel. There can be different burn rates of the same amount of fuel. For water bottle rockets, the burn rate or mass flow rate is controlled by the nozzle diameter.

Out in space without gravity, or moving sideways to gravity, we can burn the fuel slowly or quickly to get low or high thrust. A rocket can accelerate fast or slow and still reach the same speed when all the fuel is gone, just depending on the burn rate.

On Earth with gravity, we need to burn the fuel quickly to get high thrust. Rockets fighting gravity need a thrust 3 or 4 times the weight of the rocket, so that most of the energy of the fuel goes into the final speed instead of just causing the rocket to hover over the ground.

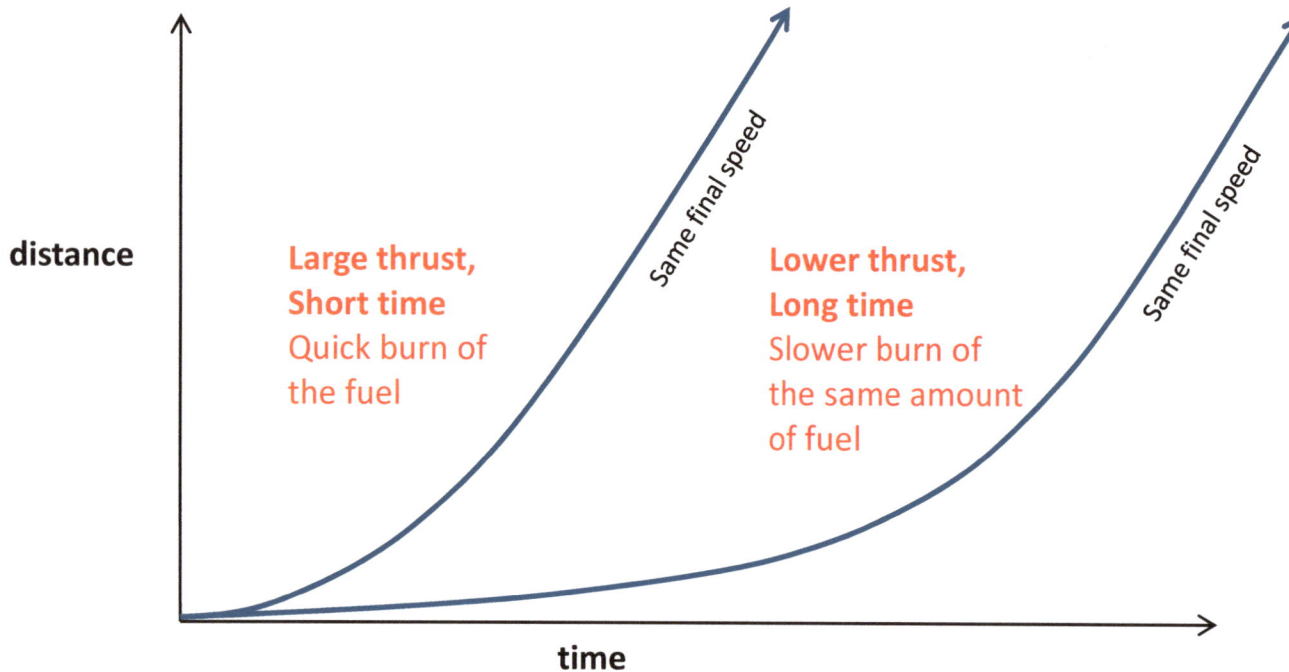

distance

Same final speed

Large thrust, Short time
Quick burn of the fuel

Lower thrust, Long time
Slower burn of the same amount of fuel

Same final speed

time

Out in space, the rocket that burns its fuel more quickly with more thrust and same amount of fuel will reach the same final rocket speed as a rocket that burns its fuel more slowly and has less instant thrust but for a longer time.

When gravity is not part of the picture, that same final velocity is true. When thrust is so large that gravity is just a small part of the picture, then the same velocity is almost true.

When gravity is similar to thrust, when the low thrust case is about the weight of the rocket, then we really want more thrust. Then burn rate is not just some secondary consideration, it is necessary to have a sizable burn rate.

Thrust is just the burn rate, but the final rocket velocity is determined by the amount of fuel, not the burn rate.

Acceleration of Water Rocket for Wide Nozzle

Bottle rockets lose water weight and accelerate even faster during the thrust. There is no throttling the thrust, or constraining the acceleration to a human friendly 3 g's.

From Newton's 2nd law, the acceleration is force/mass. So as the fuel leaves and the rocket gets lighter, then the acceleration increases.

Multi-stage space rockets also show the extra acceleration as the fuel for that stage gets used up, at the rocket gets lights. Each lower stage is the dominant mass of the rocket so the lost fuel really changes the acceleration.

"Rocket gets lighter as water is ejected, and the same exhaust force pushing against something lighter causes acceleration to ramp up."

Water Bottle Acceleration, Single Stage

Acceleration: Press = 100 psi, water fill = 34%

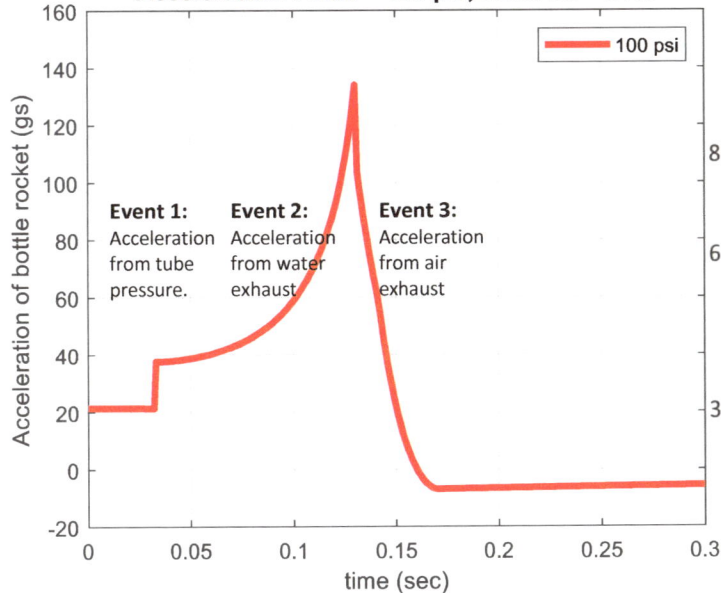

Event 1: Acceleration from tube pressure.

Event 2: Acceleration from water exhaust

Event 3: Acceleration from air exhaust

Acceleration of water rocket during thrust as water is leaving

Event 1: Tube Push Mass of rocket and fuel does not change due to tube push, so have constant acceleration.

Burnout of water: Main Engine Cut Off (MECO)

Burn-out of air

Coasting under gravity: Negative acceleration even though rocket is still traveling up: Air drag (about 2 g's) and gravity (1 g) are slowing rocket down.

What a water launch looks like, with high pressure and wide nozzle, at 50 g's.

The acceleration of the first stage of the bottle rocket has the same behavior as a space rocket first stage, although the acceleration is 10 times larger because the mass is much less.

Space Rocket Acceleration, with Multiple Stages

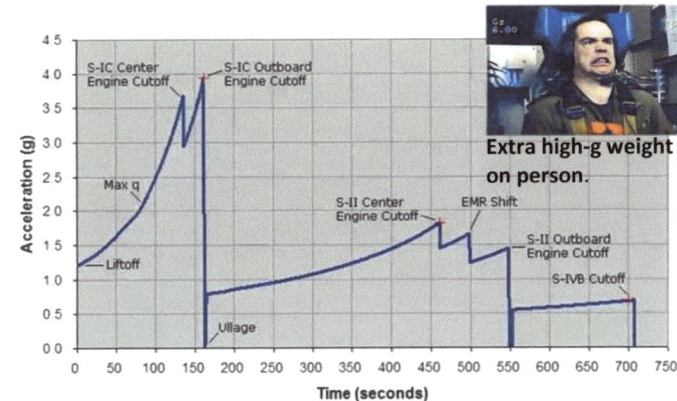

Extra high-g weight on person.

American Apollo space rocket, chemical thrust with multiple stages
Space rockets keep acceleration below 4 gs, and acceleration goes up as the fuel mass burns away.

American Apollo space rocket.

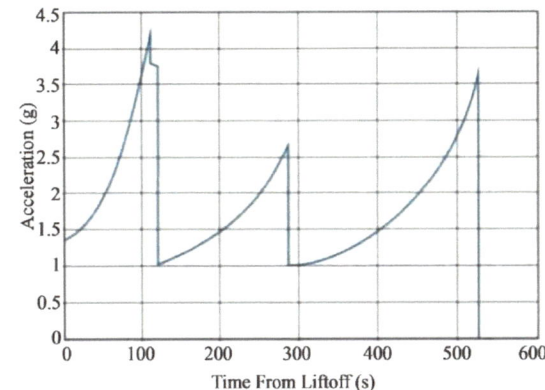

Russian Soyuz space rocket, chemical thrust with multiple stages.
The Russian rocket has larger acceleration and a shorter timeline.

Russian Soyuz space rocket

- **Force and accelerations occur for all stages of the launch: air tube, water expelled, and air expelled.**
- **Notice that the acceleration is greatest right before all the water is gone, when a nearly constant water exhaust is pushing on a much lighter rocket with barely any water in it.**

73

"Time for action! Drink some soda to get the plastic bottle, and buy a launch kit."

Get out to a field and launch a few water bottle rockets. At first, don't even worry about the design. Just buy a launcher kit and put some fins on a water bottle. Hear the bang of the launch. Hear the clunk of the landing. See if your bottle rocket can withstand the wet or broken fins and the crash of the landing.

If you want to know how high the rocket goes, compare the height to a tree or use a stop watch to measure flight time and height.

The basic fun of a launch

Fill with water

Insert PVC tube

Aim launcher

Pump air pressure

Rocket, with bottle and fins and a little top weight:
Building bottle rockets and launching them is fun, suspenseful, and engaging.

The rocket is fun to build by cutting the fins from cardboard and ripping duct tape.

Good times come from launching the rockets, pumping up the air pressure, and seeing how high or far the rocket will travel. Run some experiments without fins to see poor direction control, or with a softball weight on top to see better direction control. Run without water to see that air is an exhaust and will cause thrust too. These experiments are described in the first few appendices.

Launch tube, with seal and release mechanism:
Launch tubes enable people to pump up the bottle pressure.
Of course, you need a launch tube. At first, or in addition, just get one of the launch tubes for sale. You can build your own later. It is a project, though, to make a good seal that doesn't leak at higher pressures.

See Appendix B for bottle rocket experiments that demonstrate Newton's laws of motion and are fun.

Have fun, get wet, and learn. The basic properties of motion, like velocity and force, go hand in hand with rockets. Strong forces, greater than the weight of the rocket, are needed to have vertical launch.

Performance Concepts

These water launches are all repeatable, by refilling the water in the bottle and pumping the bicycle pump. Of course, the fins could get wet and break.

We don't need to keep buying Estes and Quest toy chemical engines and ignition fuses. That's one benefit of using water as an exhaust instead of chemicals.

Here are some of the great concepts from bottle rockets.

- **Water rockets demonstrate a lot of Space rocket concepts**. Here are some general rocket concepts. The 'Action' from pushing stuff away causes a 'Reaction'. Also, the thrust should be greater than weight to rise vertically. Also, acceleration is equal to thrust divided by mass. Also, the payload on a two stage rocket can go faster than a one stage rocket, as shown by the idea of dropping already-used weight during thrust. Also, the rocket equation predicts the rocket final velocity. Two rocket equations can be cascaded to get the final velocity of a 2nd stage.

- **Newton's laws:** If you get Newton's 2nd law, then you are doing great. That is, Newton's 2nd law says force = mass*acceleration. Newton's 3rd law says that for every force (Action) there is an equal and opposite force (Reaction).

- **Some great design concepts**: The 'Arrow Principle' works, with a heavier tip up front and fins in the back, to keep the rocket stable and going straight.

- **Exhaust velocity**: A faster exhaust velocity makes the rocket go faster and higher. Faster exhaust comes from people doing the pumping work, and pumping up the air pressure to higher pressures.

- **Nozzle diameter and thrust**: The nozzle diameter is set by the launch tube diameter, and typically you only have one launch kit and tube.   But at least we understand that narrower nozzles will have less thrust because there is less flow or exhaust at any instant. The total impulse, or total momentum change, will be the same for any nozzle diameter, because there is still the same total amount of water getting ejected at the same exhaust velocity, at either a fast flow rate or a slow flow rate (at same pressure and water fill level).

Stabilizing sideways force on bottom fins rotates the rocket back to straight, with the fins below the arrow mass.

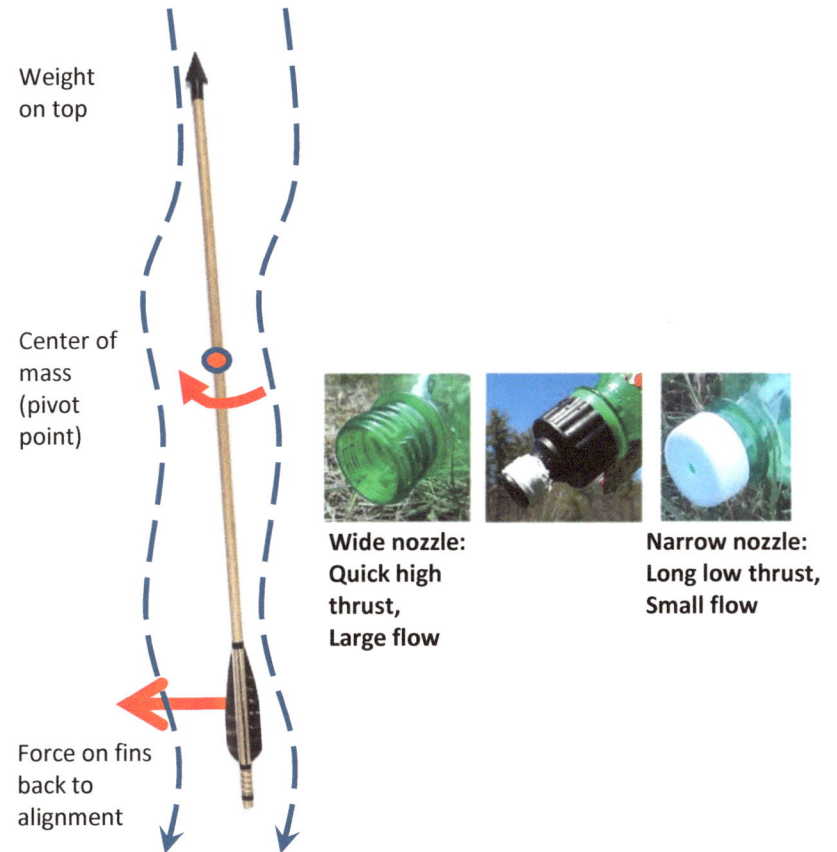

Weight on top

Center of mass (pivot point)

Force on fins back to alignment

Wide nozzle: Quick high thrust, Large flow

Narrow nozzle: Long low thrust, Small flow

The Arrow Principle: top weight and bottom fins. Top weight raises the center of mass pivot point and bottom fins give torque back to straight flight.

Have fun, get wet, and learn. The basic properties of motion, like velocity and force, go hand in hand with rockets.   Strong thrust, larger than the weight of the rocket, is needed to have vertical launch.

Some Bottle Rocket Competitions

There are many competitions and activities for water bottle rockets. One place (Colorado below) shows many of the techniques mentioned in the book, such as top weight and bottom fins. Other places (Lab in England and New Jersey below) have rockets aimed sideways toward a desired range on the ground. For a modern day flare, they add 3D printing of fins and a trash bag for a parachute.

Experiment with top or bottom weights, and parachutes.

Top weights should create more stable flight. We don't recommend any bottom weights unless you deliberately want to add instability for fun.

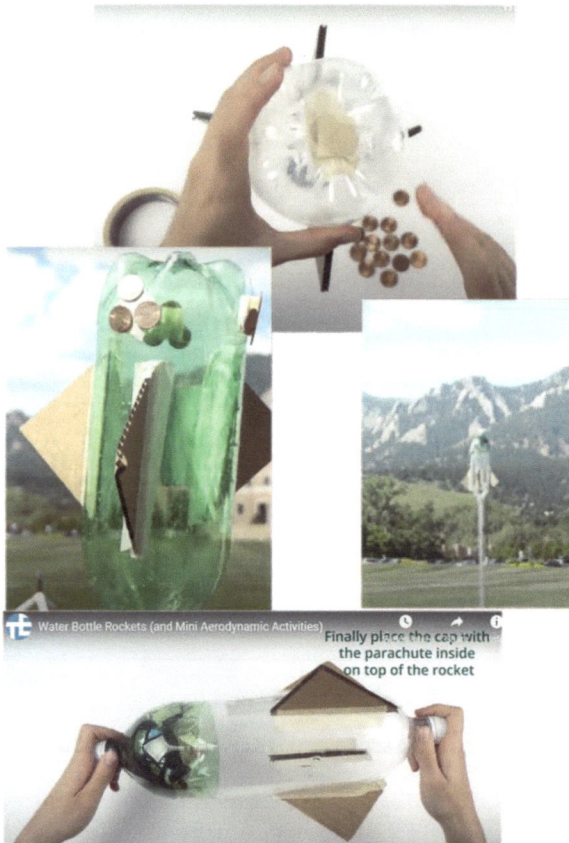

Bottle rocket description from University of Colorado, Boulder.

https://www.teachengineering.org/activities/view/ucd_bottlerockets_activity1

Experiment with aiming rocket for a specific range.

Bottle rockets don't need to be aimed straight up. If can aim for a target on the ground some distance away, the challenge is to have a stable flight, the right pressure, and the right amount of water.

Different tilt angles require different pressures for the same specific range. A more vertical launch will require more speed and altitude with more loft or hang time.

National Physical Laboratory in England

https://www.npl.co.uk/water-rockets

Experiment by making bottle vehicle with fins and caps or nose cone made from 3D printers and laser cutters.

3D printing of plastic is not known for being strong, so your cap might shatter on impact.

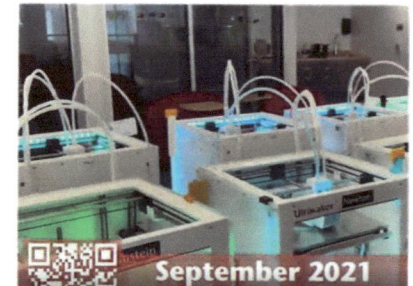

New Jersey Institute of Technology

https://www.njitmakerspace.com/rocket

See what other people are doing with bottle rockets on the internet.

A Blast with All Things Water

Water can be powerful like a breaking wave on the beach slamming your body. Water can be given high pressure with a turbo pump or with pressurized air above it.

For the bottle rocket, compared to chemical rockets, we don't need to keep buying Estes and Quest toy chemical engines and ignition fuses. But of course we'll also keep buying chemical engines because they are exciting too. Water bottle rockets don't require trips to the store. We just need sweat and muscle for the bicycle pump.

Let's see common things that use thrust from water.

Lakes can provide a continuous source of water for a jet pack. Some missile launches from submarines get their initial thrust from steam, rather than exhaust gases from a chemical reaction.

Interestingly, the exhaust from a space rocket with hydrogen fuel and liquid oxygen oxidizer is hot water. This exhaust water is very hot so it has much higher exhaust velocity than from a bottle rocket.

"You'll see water rocket exhaust thrust in many places, like squids, fire hoses, and even chemical products of space rockets."

Fire hose:
About 200 pound-force, where two firefighters are needed to aim the spray.

Launch of Very High Pressure Water Rocket:
500 psi hose using electric pump into PVC rocket, more than 500 pound-force thrust.

Evel Knievel, with Steam rocket:
Evel Knievel crosses a canyon using hot steam, more than 1000 pound-force thrust.

2nd stage of rockets, with water exhaust:
Combining liquid H and O to get water or hot steam for thrust, more than 100,000 pound-force thrust.

Squids in the ocean

Water Jet pack:
The water is blasting out fast, and the pressure comes from a pump floating on the water.

Rocket Man in 2018, with steam rocket:
Rocket using steam power to blast up 2000 feet.

Poseidon rocket, with steam ejection:
A submarine rocket uses steam to eject from the launch tube, before chemical rocket kicks in.

Bigger size, higher performance

Water thrust is used for squids, for tourist jet packs. Water exhaust is actually the most efficient, highest energy, exhaust for rockets in space, combining hydrogen and oxygen to make really hot steam with an exhaust velocity of 4400m/s.

The main roadblock to getting started is the launcher to pump the bottle up and then release, so here is one set of reviews of launch kits:

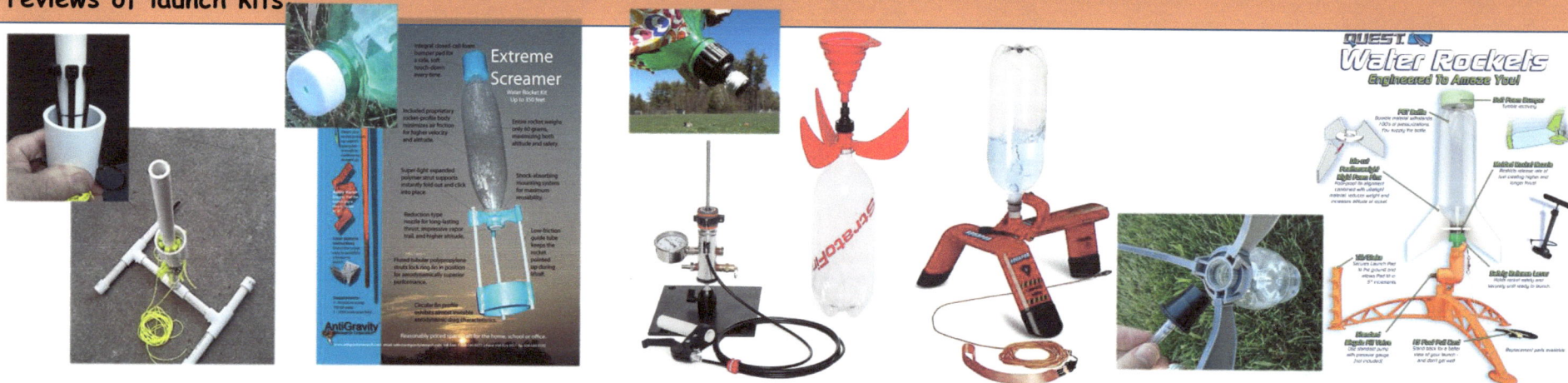

Soda Bottle Water Rocket Launcher
Toy Do It Yourself Kit. Prefabricated Parts, 10 to 15 Minute Assembly, from PVC Creations.

Score 5:
Cost: $30
Psi: 100 psi

Pro:
- This pre-made PVC kit is basically a professionally designed home-built PVC design, with a bump for a seal, with only one trip to the hardware store to get PVC glue.
- Get high thrust launch, gets to 100 psi.
- The launcher is low cost, reliable, high pressure, and the students can all build separate rockets.

Con:
- Need to assemble tubes with PVC glue.

'Anti-gravity' Water Bottle Rockets, 1-stage and 2-stage

Score: 4
Cost: $100
Psi: 100 psi

Pro:
- Very easy to set up.
- Can demonstrate 2-stage rockets.
- Can control air pressure by disconnecting the hose from the bicycle pump.
- Already designed for student demonstrations. Comes as a kit intended for demonstration of rocket physics, with different nozzle sizes.
- Various kits available. Available as a 1-stage or 2-stage rocket.
- The bottle cap nozzles can be attached to any student built rocket as well as the provided rockets.

Con:
- Fins are fragile and are meant to flop off during landing. Just need patience putting fins back on each time.
- Can not launch without fins, because the fins support the rocket during launch. But this is minor.

Relationshipware StratoLauncher IV Deluxe Tilting Water Rocket Launcher + StratoFins Kit

Score: 4
Cost: $80
Psi: 100 psi

Pro:
- Very reliable and high pressure, when use plumbers tape.
- Gets to the highest pressure, limited by the bicycle pump.

Con:
- Not a long launch tube to guide the rocket, so rocket can fly sideways.
- Need to assemble with Teflon tape.

AquaPod Water Bottle Rocket

Score: 3:
Cost: $30
Psi: 30 psi

Pro:
- Very easy to set up.
- Good for a quick launch.

Con:
- Can not get higher pressures above 30psi.

'Science in Action' Water Bottle Rocket

Score: 3:
Cost: $30
Psi: 20 psi

Pro:
- Very easy to set up.
- Good for a quick launch.
- Good for really young kids.

Con:
- Can not get higher pressures above 20psi.

Quest Water Bottle Rocket

Score: 3:
Cost: $30
Psi: 50 psi

Pro:
- Very easy to set up.
- Good for a quick launch.
- Comes in kits for classrooms.

Con:
- Fragile.

There is no wrong answer for launch kits. Get what is available.

Air Pressure and Action and Reaction

This book can be used as a source to create rocket demonstrations. Water rockets are great, because the water bottle rockets demonstrate the basic behavior of chemical rockets. For example, more exhaust velocity and more fuel (higher pressure air and water) and more burn time will give a faster final rocket velocity.

Try some of the experiments in the book, like
- Try a high pressure bottle rocket with and without water, and compare the height of this air blast versus this water blast.
  - Hey, any exhaust, air or water, causes thrust.
- Try different water levels to see the rocket altitude.
  - Too little water has too little thrust duration, and
  - Too much water has too little compressed air.
- Try different weights on the top of the rocket to see if the rocket is more stable and flies straighter.
  - Rocket should fly straighter like darts, in a dartboard game.
- Try different nozzle diameters to see the different forces and accelerations.
  - Thrust is larger with more flow rate using larger diameter nozzles.

Building the rocket:

The budding rocket engineers can be responsible for the rockets, or a few rockets can be provided by the commercial kits or the teacher.

Read Chapter 5 for instructions for students to build the rockets. You'll need poster board, duct tape, and scissors.

In this appendix are experiments that show how to get straight flights and high altitude, using the 'Arrow Principle' with bottom fins and top weight.

Building the launcher:

The parent or teacher is responsible for building or buying the launch pad. For example, the plastic 'PVC Creations' kit just needs glue and assembly, and will fit 2 liter bottles. The metal 'Relationshipware' kit has a nozzle cap that can be screwed onto any 2 liter bottle. The 'Anti-gravity' kit does not use a launcher, and this kit also has bottle caps with the narrow hole that can be screwed onto any 2 liter bottle.

Try whatever tickles your fancy. Water bottle rockets are all fun and games.

Air Pressure and Common Usage

Look around at all the uses of high pressure air.   - car and bicycle tires, water fountains, tools, and compressed air for scuba diving.

1. **What are some common uses of air pressure?**
    *   Tire pressure,
    *   Water pressure in houses and buildings
    *   Air-powered tools
    *   Swimming underwater with scuba tanks
    *   Compressed air with diesel fuel vapor heats and self ignites in diesel engines.

Tire pressure
Air pressure is a great shock absorber, and lets the rubber hug the road.

Air pressure pushing water fountain
Air pressure is constantly pushing on the water tank, so the fountain flows without turning the air compressor back on.

Tools in car garage and machine shops:
Pneumatic tools spin using compressed air. A lot of torque and instant power is available with the high pressure turning a turbine-style motor, or the air can push a piston motor.

SCUBA Tanks:
Highly compressed air (>3000 psi) keeps scuba divers breathing.
The air pressure needs to be larger than the water pressure, and hold a lot of breathable air.

Reliable trucks:

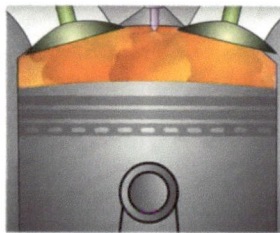

Explosion inside hot compressed air:
compressed gas, fuel, and self ignition

Diesel engines:
Diesel fuel is injected into hot compressed air in the cylinder and vaporized and self ignites with the oxygen.
This is not just compressed air, because a chemical reaction happens, but the compressed air is needed to create the heat and ignite the diesel.

There are plenty of air pressure toys out there to play with, besides practical uses of air pressure.

Air Pressure and Action and Reaction

Before playing with your own homemade water bottle rockets, introduce the whole idea of rockets by using all the toy rockets available. The most important concept to get out of this rocket demonstration is 'Action and Reaction', which allows a rocket to get thrust and go up against gravity.
All these examples below show that air, water, or bullets create an exhaust, so there is thrust.

2. **Examples of thrust and kickback, or action and reaction?**

- <u>Water hose</u>: Can you feel any kickback when you quickly turn on a water hose?

- <u>Water faucet</u>: Can you see any kickback from a household sink faucet when turn on water quickly, especially after the filter is removed to allow more water flow?
  - Anytime something is getting ejected, there is Action and Reaction, or a backwards force. So both the hose and the faucet have a kickback.

Video of exhaust speed

Water exhaust

- <u>Balloons</u>: Let kids play with inflated balloons, and release them and fly away.
- <u>Stomp rockets</u>: Let air pressure blast off a tube rocket.

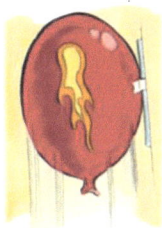

Rocket balloon sliding on a string

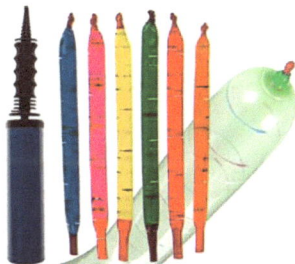

Rocket balloon that squiggles around the back yard.

Stomp rocket

- <u>Pellet guns</u>: If the kids are old enough, let them shoot pellet guns and see the kickback as the barrel pivots up after the pellet is shot.

- <u>Boats</u>: Push off the end of a canoe and see the canoe go the other direction. You are the 'exhaust'.

  It is hard to step out of a canoe, without the canoe moving backward out from under you.
  ...be ready to get wet.
  This backward push is an example of action and reaction, or conservation of momentum.

Pushing off a canoe shows Action and Reaction

There are plenty of air pressure toys out there to play with, besides practical uses of air pressure.

'Sight only' means no fancy measurements, although maybe a stop watch can be used (or just count the seconds). Flight time will give the altitude. You can also just use your eyes and see if a rocket travels straighter with good fins on the back.

3. Do fins help to stabilize flight, to make rocket go straight up?

- Remove feathers from an arrow or dart and see if the remaining shaft travels as straight.
- Launch a bottle rocket without fins and see if the bare bottle travels as straight.
  - A bottle without fins can work if there is enough top weight. Then the bottom of the bottle gives enough sideways drag when tilted to rotate and right the bottle and keep the flight straight. This is the same as dragging long sticks, to keep fireworks stable. Anything that creates the sideways drag at the bottom to right the rocket works.

Fins	Height (flight time)	Veer sideways		No Fins	Height (flight time)	Veer sideways
Trial 1				Trial 1		
Trial 2				Trial 2		

Compare fins versus no fins for height and straight flights.

Stick drag for stability

4. Should fins be on the bottom or top?

- Put fins on the top of the bottle and see if the rocket becomes completely unstable.
  - The sideways drag needs to be below the center of mass in order to allow stability. The sideways drag is at a location called the 'center of pressure', which needs to be below or behind the center of mass to right the rocket back to a straight flight.

Fins on bottom	straight	Veer sideways	flipping	Fins on top	straight	Veer sideways	flipping
Trial 1				Trial 1			
Trial 2				Trial 2			

Use top and bottom fins?

- Do something completely wrong and get on a paddle board on a lake with the board backwards and the fin in front. The paddler has a very difficult job to go straight. Any drag on the front fin always pushes the front to turn.

5. Are smaller or larger fins better to stabilize flight?

- Do larger fins work better at lower rocket speeds (lower pump pressure)?
  - Recall that darts have large fins in back because darts fly slowly and need to create enough sideways drag during tilts at these slow speeds.
  - The question is whether a bottle rocket is traveling fast or slow. Because the bottle rocket is still accelerating for the first 10 feet, and not at maximum speed, then larger fins will help keep the rocket straight during the start of launch at slower speeds.

Slow darts that fly at slow speeds have big fins for straight flights, in addition to weight at the top.

Larger fins on Bottle	Height (flight time)	Veer sideways		Smaller fins on Bottle	Height (flight time)	Veer sideways
Trial 1				Trial 1		
Trial 2				Trial 2		

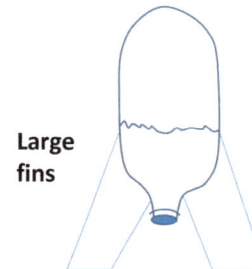

Large fins

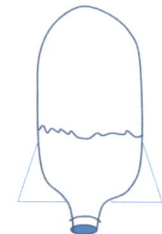

Small fins

Compare large and small fins for height and straight flights.

You don't need to record anything, just observe. If a rocket spins out of control, you know whatever you did didn't improve stability.

Stopwatch (cell phone timer) Things to Try: 'Arrow Principle' and Air Drag

6. **Does top weight improve stability and flight time?**
   - Add some top weight.   If the flight time gets longer or stays the same, that means that the rocket is blasting through air easier from Littleton's rule.   Make the bottle heavy enough with slight added weight that air drag doesn't slow the rocket down as much.
   - **Use different weights on top of rocket, and compare flight times for a water rocket.   Use consistent pressure and water level.**
       - Make a plot of dry weight versus time or altitude, using many different dry weights and different launches.
       - Too light dry weight and the air drag will easily slow down the lighter rocket. Too much dry weight and the rocket thrust does not accelerate the heavy rocket to as fast a speed and the rocket does not go as high.   So there is an optimal weight for the dry weight, which is just a little more than the bottle weight. Air drag is already larger than gravity at the beginning, so make the weight larger.
       - As shown in the measurements below, the flight time does not change until the top weight is more than the bottle weight, because air drag is less important when the same rocket is heavier.

Added weight	Height (flight time)
4 quarters	
8 quarters	
12 quarters	
16 quarters	
20 quarters	
24 quarters	

Record flight time for many different top weights.

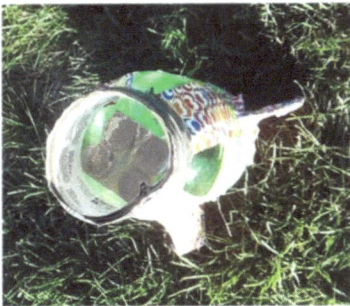
Top pouch for top weight

flight time (sec): 15% water, 40psi

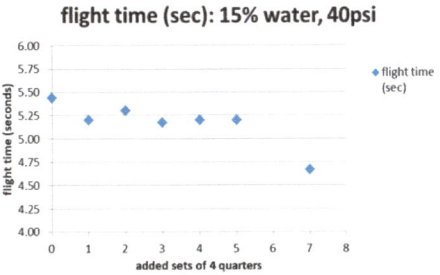
Flight time versus top weight

Groups of 4 quarters for top weight

7. **Does bottom weight not help to stabilize flight?**
   - Try bottom weight, and see if the bottle just flips around during launch.   Bottom weight should make the bottle rocket unstable.
       - Sometimes to see if something has an impact, it is useful to do the exact opposite and see if there is a total breakdown in stability.   And, yes, the bottom weight should cause a total breakdown in stability. The sideways drag when the bottle starts to tilt will only cause the bottle to keep tilting, because the fin or sideways drag is above the center of mass.   Air drag is pushing sideways on the nose.   In rocket or airplane speak, the 'center of pressure' is above the 'center of mass'.

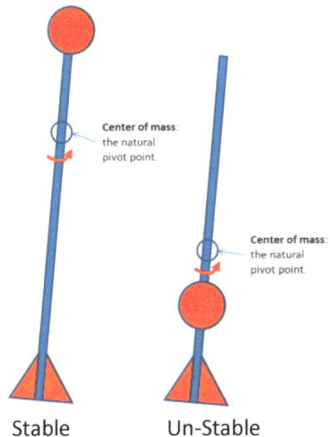
Center of mass: the natural pivot point.

Center of mass: the natural pivot point.

Stable   Un-Stable

Keep center of mass up high for straight flight

Experiments that show air drag:

8. **Does air drag matter?**
   - Compare flight time of a fat and skinny bottle, with the same pressure and water and dry weight.
   - Add a cardboard disk around the bottle rocket, and see if the flight time is less.
   - Compare a flat top versus a cone top.   Which stays up longer?

rocket	Height (flight time)
Skinny, 2 liter	
Wide, 2 liter	

Altitude: Press = 100 psi, water fill = 34%

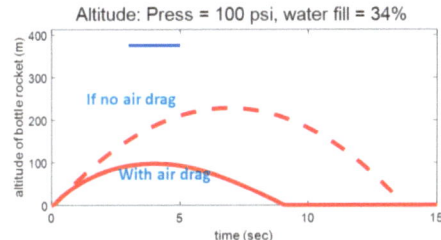
If no air drag

With air drag

Air drag dramatically slows down rocket

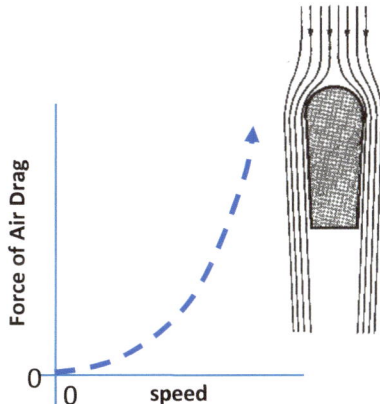
Force of Air Drag

speed

Air drag versus speed

Stopwatch (cell phone timer) Things to Try: Thrust

Here are experiments that show thrust, or demonstrate Action and Reaction: What's the best water height? Can all nozzle sizes lift the water? How much does more pressure help?

9. Does water beat air?

- What has more flight time and altitude? Air or water rockets at same pressure?
  - The water rocket has a longer exhaust time. The water rocket also has both the water thrust and the air thrust afterwards as the high pressure air escapes after the water is gone. So the water rocket should go higher for the same air pressure.

Case all at 50psi	Height (flight time)
Air only	
10% water	
20% water	
40% water	
60% water	
80% water	

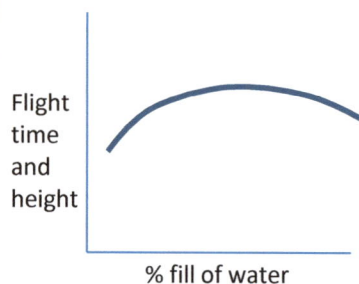

Flight time and height vs % fill of water

10. Can there be too much water, and not enough compressed air to push the water out? (use data above)

- Plot flight time versus water fill level. Above 40% fill is overfill, and the rocket should not go as high because there is not enough air to push the water out.
  - If you only leave 1 inch of top space for the compressed air in the bottle, the rocket is not going to have much air and energy and will not have any sustained pressure on the water as the air expands. There is less energy in the small volume of compressed air. The optimal water fill fraction is less than 40% for wide nozzles, and less than 20% for narrow nozzles. The water has weight and the nozzle size needs to be wide enough to allow enough water to flow out quickly to create more thrust than water weight.

11. Does the rocket fly more easily to the side or more easily deviate or tilt uncontrollably using more water? (use data above)

- With more water, the rocket will be heavier and the velocity will be less, so the fins shouldn't work as well with less air flow over their surface. Also, with more water, the center of mass during the launch is lower or near the bottom, making the rocket less stable.

12. Does more pressure make the rocket go faster?

- Plot flight time versus pressure. Flight time increases with pressure, where altitude goes as flight time squared (Littlewood's rule).
- Plot acceleration versus pressure, using many videos at different pressures. The acceleration should be proportional to the pressure.

Case all at 20% water	Height (flight time)
30 psi	
50 psi	
70 psi	
90 psi	

Case all at 20% water	Acceleration from videos
30 psi	
50 psi	
70 psi	
90 psi	

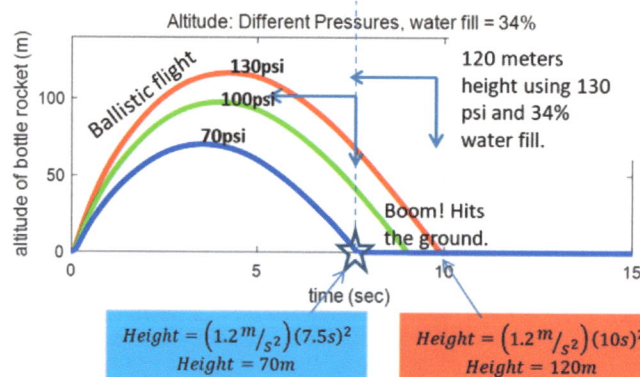

Altitude: Different Pressures, water fill = 34%

120 meters height using 130 psi and 34% water fill.

$$Height = \left(1.2\,{}^{m}/_{s^2}\right)(7.5s)^2$$
$$Height = 70m$$

$$Height = \left(1.2\,{}^{m}/_{s^2}\right)(10s)^2$$
$$Height = 120m$$

Height versus time, with air drag

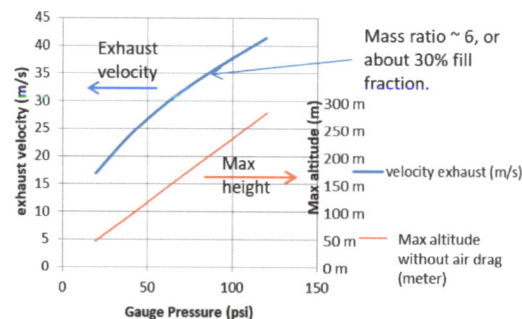

Velocity Exhaust (m/s) and Max Height (m)

Mass ratio ~ 6, or about 30% fill fraction.

Exhaust velocity and max height versus pressure, ignoring air drag

Again, a stopwatch will tell you how high the rocket flew. Use it to see if there is a best water fill value, or to see how much pressure improves height.

Video (cell phone) Things to Try: Thrust and Videos

Here are experiments that measure thrust and acceleration using a video camera. We can measure acceleration and use Newton's third law to get thrust, or we can get a scale and measure thrust.

13. Want to know the thrust value?

- Measure the acceleration using a video camera and video editor, by recording the altitude versus time. Then work backwards from acceleration to get the force, using F=ma to get the force.
  - A video will show the acceleration of the rocket. You will need to have a video editor to stop the video frame by frame to get the location and time of the rocket in each frame. Also, it helps to use a faster frame rate, more than 30 frames per second, to capture the motion of the fast moving rocket (for the wide nozzle thrust). Most cameras have a 'slow motion' setting which has a faster frame rate.

  Use video to get the average acceleration:
  1. Get fastest or final velocity from the video, by measuring distance change between a frame after thrust is nearly over or exhausted.
  2. Note the total time to get to that final velocity, from start of thrust
  3. Average acceleration = final velocity/total time

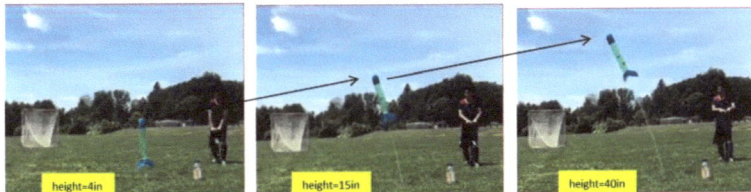

height=4in
T=0 sec

height=15in
T=0.125 sec

height=40in
T=0.250 sec

Launch of narrow nozzle bottle rocket

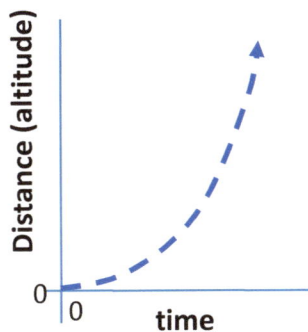

Distance versus time for a constant acceleration

The distance keeps increasing at a faster rate versus time when there is a constant acceleration (speed is increasing each second):

If constant acceleration:

$$height = \frac{1}{2} acceleration * time^2$$

Record the height versus time, and estimate the acceleration that will best fit the height.

14. Measure thrust from a hose using a scale or flow of water:

- **Scale**: Blast the water stream from a hose onto a scale and record the thrust.

$$F_{exhaust} = (Read\ Scale)$$

water

scale

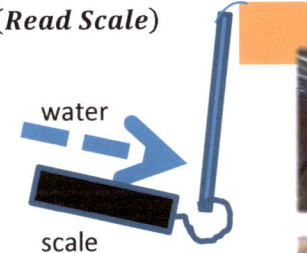

The scale is wrapped in a sandwich bag to keep the scale dry.

Direct measurement of thrust: fish scale or letter scale

- **Flow of water**: Use the thrust equation to estimate the thrust. Measure the water flow rate by filling a bullet in a known time, to get change in mass over change in time. Measure the exhaust velocity with a video to see how fast the stream of water from the hose moves over a few meters.

$$F_{rocket} = v_{exhaust} \frac{change\ in\ mass}{change\ in\ time}$$

3 meters

Speed: Measure speed of ejected water using high speed video on camera, along with video software.

Mass rate: Measure fill time of water into a container

Exhaust velocity and flow rate measurements

A video and looking at each frame will give the acceleration and thrust. You can also directly measure thrust in backyard experiments.

Here are 5 techniques to build a homemade launcher, the hard way using materials from scratch.

"Deep down in their gut, people get this. The launcher needs a tight seal, and need a pipe going above the water level inside the bottle."

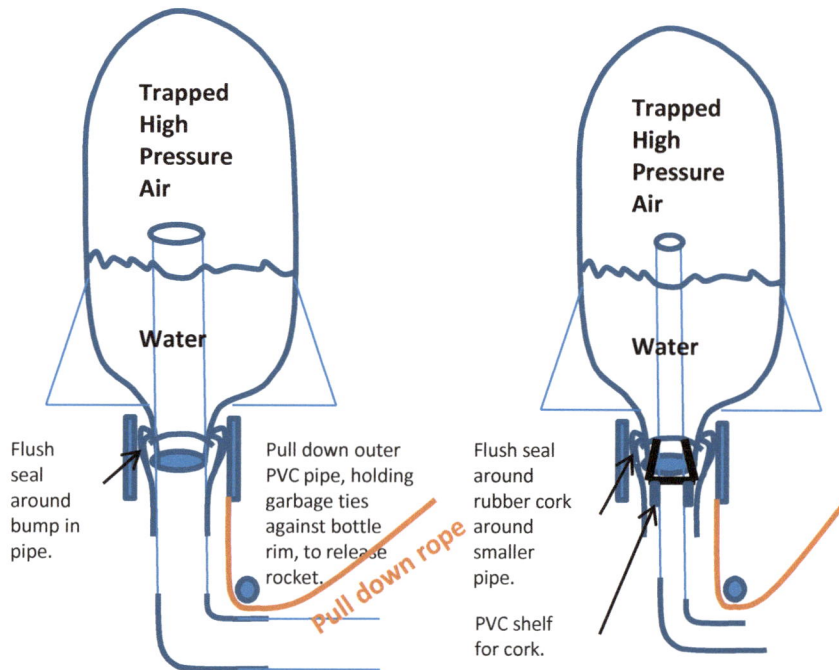

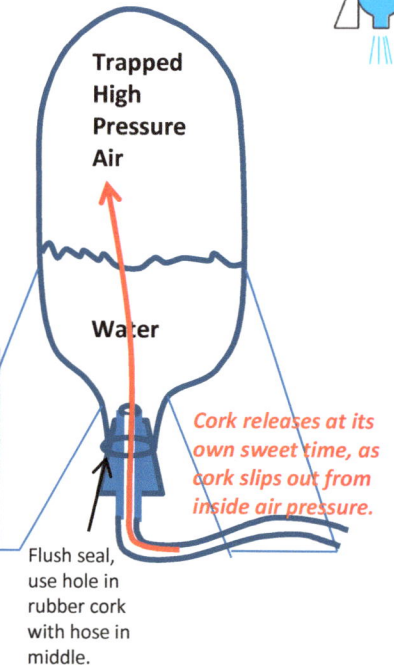

Trapped High Pressure Air

Water

Flush seal around bump in pipe.

Pull down outer PVC pipe, holding garbage ties against bottle rim, to release rocket.

Pull down rope

Trapped High Pressure Air

Water

Flush seal around rubber cork around smaller pipe.

PVC shelf for cork.

Trapped High Pressure Air

Water

Cork releases at its own sweet time, as cork slips out from inside air pressure.

Flush seal, use hole in rubber cork with hose in middle.

Technique 1:
Melt a bump in PVC pipe using heat, for bottle seal. Use slide-able clamp to hold garbage ties over rim of bottle. The rocket launches when the clamp is pulled down with the rope.

Technique 2: Insert a rubber cork around PVC pipe, for bottle seal. Use a hose clamp to hold garbage ties over rim of bottle. The rocket launches when the clamp is pulled down with the rope.

Technique 3: Tighter push-down on bump. The rocket launches when the latch is released.

Technique 4: Hold down rocket against seal with hands. The rocket launches when hands are released.

Technique 5: Rubber Cork, for bottle seal, where the cork releases by itself when pressure is pushing enough, and the rocket launches by itself.

Build this PVC launch pad yourself or buy the same concept pre-made kit.

Case 1: Homemade 'Hard Way' Launch Pads

Technique 1: Melt a bump in PVC pipe using heat, for bottle seal. Use slide-able clamp to hold garbage ties over rim of bottle. The rocket launches when the clamp is pulled down with the rope.

"Deep down in their gut, people get this. The launcher needs a tight seal, and need a pipe going above the water level inside the bottle."

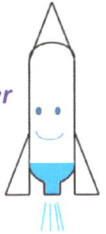

Bottle rockets are just pressurized bottle pushing out water. Bicycle pumps can get up to 100 psi, which is a safe pressure for plastic bottles.

Technique 1: Melt a bump in PVC launch tubes, and you have control over the release pressure because garbage ties are holding the rocket down until released by pulling the rope.
A candle or hair dryer can be used to soften the plastic PVC pipe.

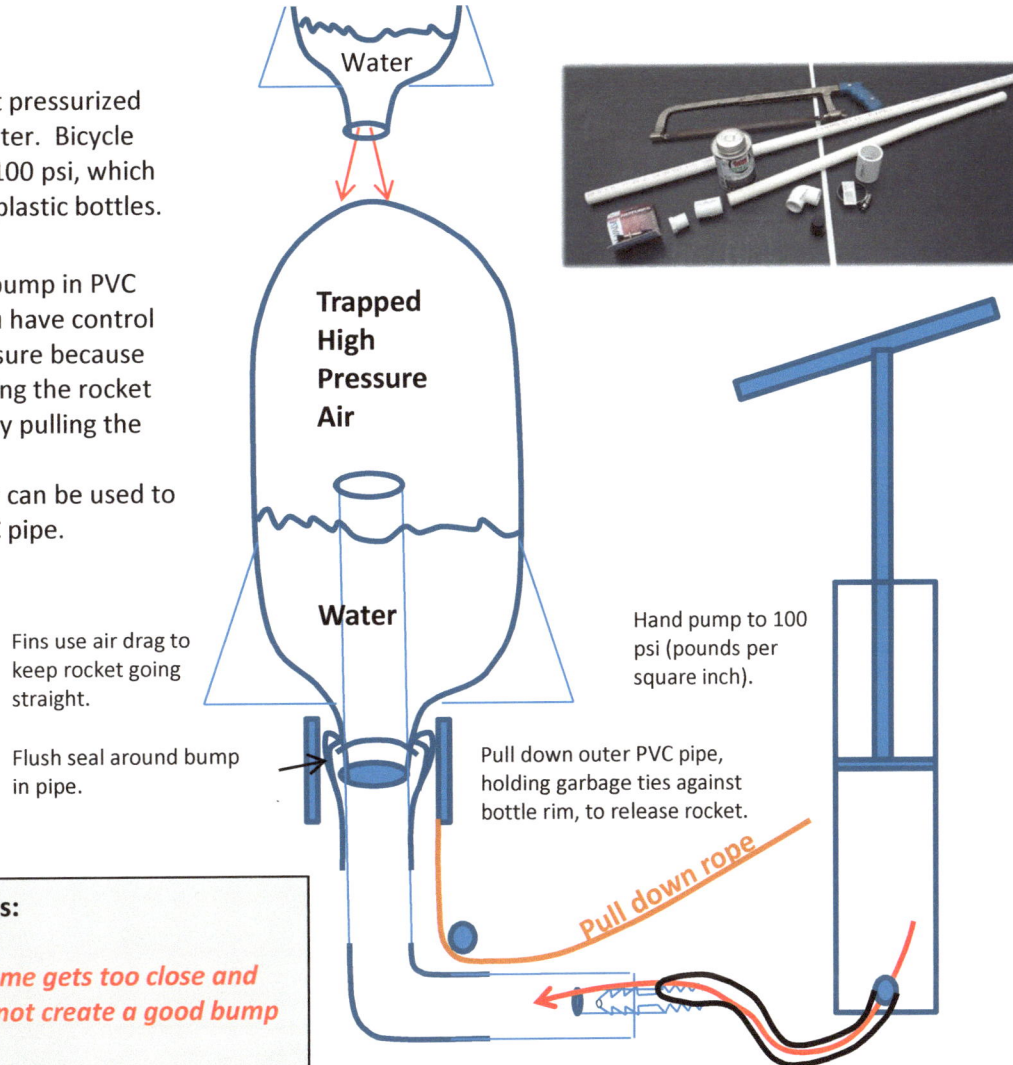

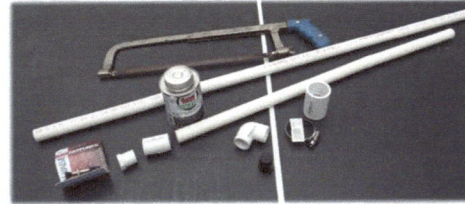

Pre-made Kit versions, also with PVC bump for seal

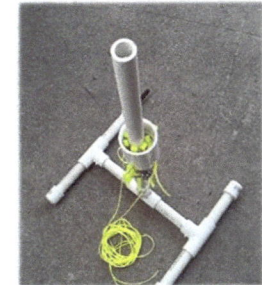

Soda Bottle Water Rocket Launcher, from PVC Creations.

Water

Trapped High Pressure Air

Water

Fins use air drag to keep rocket going straight.

Flush seal around bump in pipe.

Pull down outer PVC pipe, holding garbage ties against bottle rim, to release rocket.

Hand pump to 100 psi (pounds per square inch).

Pull down rope

Rocket Launcher for Water and Soda Bottles.

Hardware store follies:
1st trip: get PVC pipes
...let's assume the flame gets too close and burns PVC so you do not create a good bump the first time.
2nd trip: buy more PVC pipe

Build this PVC launch pad yourself or buy the same concept pre-made kit.

Homemade 'Hard Way' Launch Pads

Technique 1 continued: Bump in PVC pipe using heat, for bottle seal

Here is how you make a launcher option 1, using standard PVC piping.

1. Go to the hardware store and get PVC piping of the exact diameter as the opening of the plastic bottle you plan to use. Not all bottles have the same diameter opening, so bring the bottle with you to the hardware store.

2. The bump in the PVC piping is easy to make because the PVC pipe melts so easily. Put a clean flame under the pipe, while rotating the pipe for uniform heating. When the pipe gets soft, then press the two ends together a little bit, and a smooth bump forms. The outer diameter of the PVC needs to be exactly the same diameter as the bottleneck. This tight fit requires bringing the bottle to a hardware store that has a lot of variety of pipe diameters.

Hardware:
- PVC straight pipe
- PVC 90 degree elbow
- PVC caps
- PVC glue
- Candle or hair dryer
- Garbage ties

Step 1: Create bump in PVC pipe using heat from flame

Trash bag ties

Bump in PVC pipe

Step 2: Position garbage ties at right height.

Step 3: Hold ties on rim of bottle using pipe clamp and large pipe piece.

Snug outer pipe clamp to hold ties in.

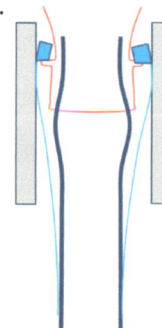

Step 4: Release ties by pulling on rope and releasing the ties, to launch rocket.

Pull down the outer pipe to release rocket.

Technique:
- Get PVC diameter exactly equal to bottleneck opening
- Can use a candle to heat PVC, and rotate PVC while pressing two sides together to make a smooth bump. PVC plastic is very soft when gently heated.

Design Limits:
- The seal depends on the smoothness of the bump, although Teflon tape may reduce leaks under pressure.

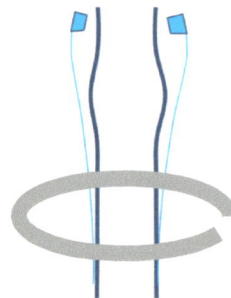

Option to pull ties down with a pipe clamp, to get exact stop to hold bottle rim down tightly.

Keep the flame enough away from the plastic so you don't burn the pipe. Squeeze the pipe together to create a bump.

Technique 2: Insert a rubber cork around PVC pipe, for bottle seal. Use a hose clamp to hold garbage ties over rim of bottle. The rocket launches when the clamp is pulled down with the rope.

"Deep down in their gut, people get this. Just need a tight seal, and need pipe going above the water level inside the bottle."

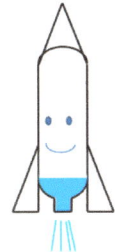

Bottle rockets are just pressurized bottle pushing out water. Bicycle pumps can get up to 100 psi, which is a safe pressure for plastic bottles.

Technique 2: Drill a hole in a rubber cork and slide the cork over the PVC launch tube. You have control over the release pressure because garbage ties are holding the rocket down until released by pulling the rope. Make a PVC 'shelf' for the rubber cork to rest against, to stop the cork from sliding down.

Water

Trapped High Pressure Air

Water

Hand pump to 100 psi (pounds per square inch).

Fins use air drag to keep rocket going straight.

Flush seal around rubber cork around smaller pipe.

PVC shelf for cork.

Pull down outer PVC pipe, holding garbage ties against bottle rim, to release rocket.

Pull down rope

Hardware store follies:
1st trip: get PVC pipes and rubber cork
...rubber cork ripped apart by drill bit and forgot PVC step below cork.
2nd trip: buy more PVC pipe and cork

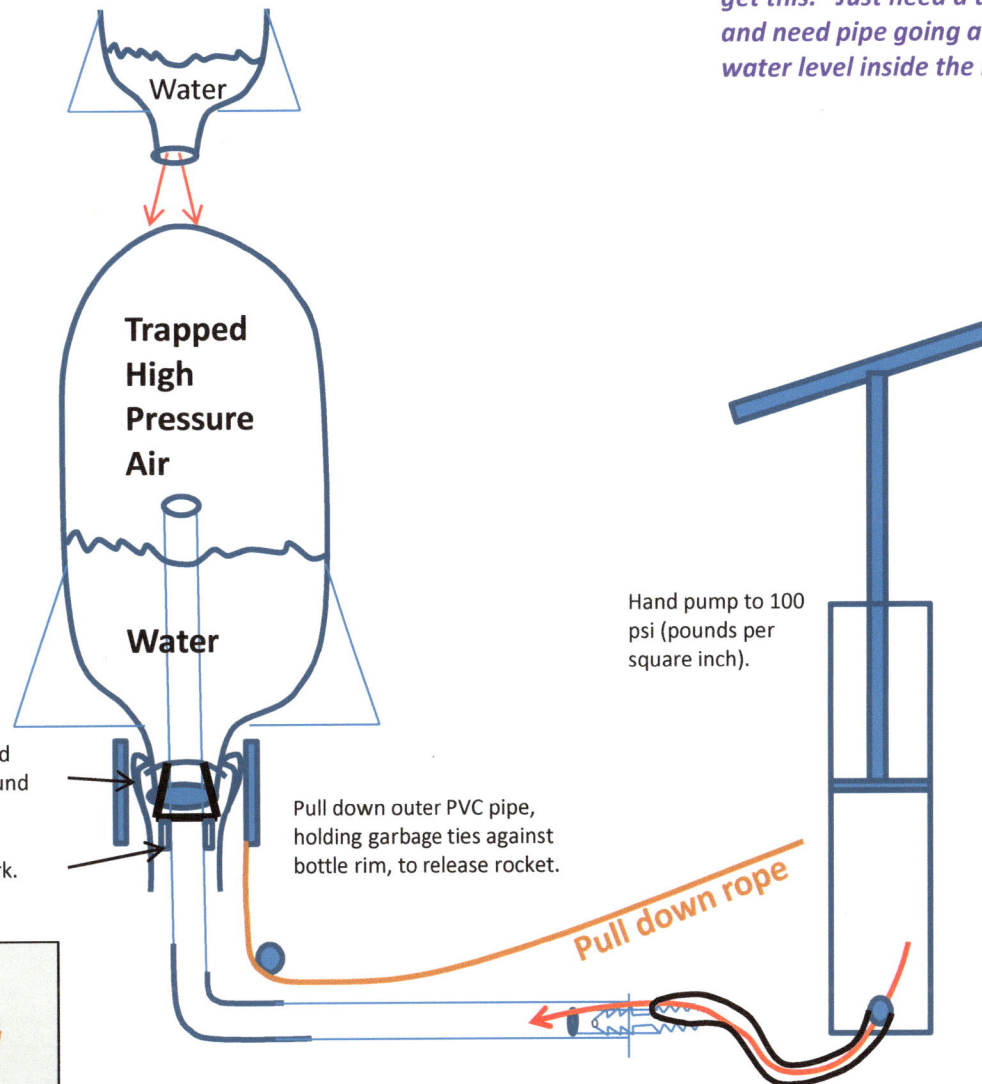

Rubber corks give more slop to get a good seal.

Homemade 'Hard Way' Launch Pads

Technique 2 continued: Rubber Cork, for bottle seal

Step 1: Drill out rubber cork and slide over PVC pipe.

Step 2: Position garbage ties at right height.

Step 3: Hold ties on rim of bottle using large pipe piece.

Trash bag ties

Rubber cork

Outer sleeve for stopper

Hold ties on with snug outer pipe or hose clamp.

Technique:
- PVC pipe diameter smaller than bottleneck opening.
- Slide rubber cork over pipe, after drill hole through it.
- Freeze the rubber cork before drilling, if having trouble drilling

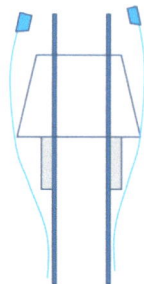

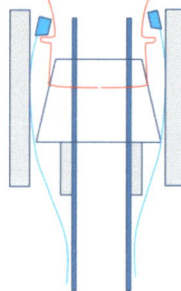

Design limits:
- The water started leaking out above 80 psi pressure. Probably the ends of the ties need to lowered and press down tighter on bottle. An improvement is to use a pipe clamp holding the ends of the ties which is adjustable.

PVC tube diameter :
- Find a cork that fits the bottle, and can use a much smaller diameter PVC tube through the cork.

When you use a rubber cork around the PVC pipe, then the pipe does not need to match the bottleneck diameter exactly.

Case 3: Launcher Technique 3: Tighter push-down on bump. The rocket launches when the latch is released.

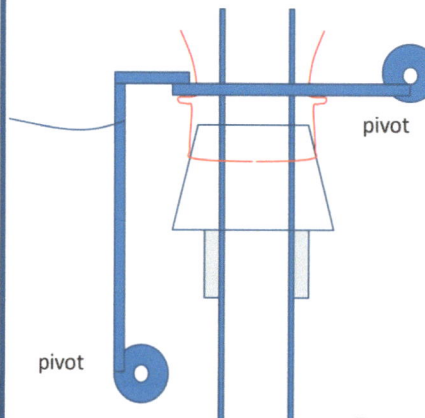

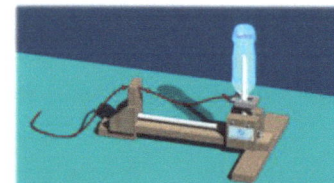

pivot

pivot

Fancy clamp that holds down rim of bottleneck.

This clamp is the most professional launcher, if you want to get really high pressures. Have a latch holding down the pressure plate. Yank the latch away and the pressure plate swings up.

Case 4: Launcher Technique 4: Hold down rocket against seal with hands. The rocket launches when hands are released.

Basic option:
- Simple hand hold against rubber cork, or PVC bump.
- Easier with two people

Design limits:
- Can get to about 40 psi without water leaking.

Working as a team to pump and hold rocket down

Case 5: Homemade 'Hard Way' Launch Pads

"Just need a tight seal, so rocket can build up some pressure before self-launching."

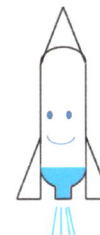

it's not
rocket surgery

Bottle rockets are just pressurized bottle pushing out water. Bicycle pumps can get up to 100psi, which is a safe pressure for plastic bottles.

Technique 5:
Pressure fit cork, no piping. Rocket launches whenever the cork slips out.

Water

Trapped High Pressure Air

Push more air above water, up to 100 psi.

Water

Fins to use air drag to keep rocket going straight.

Flush seal, use hole in rubber cork with hose in middle.

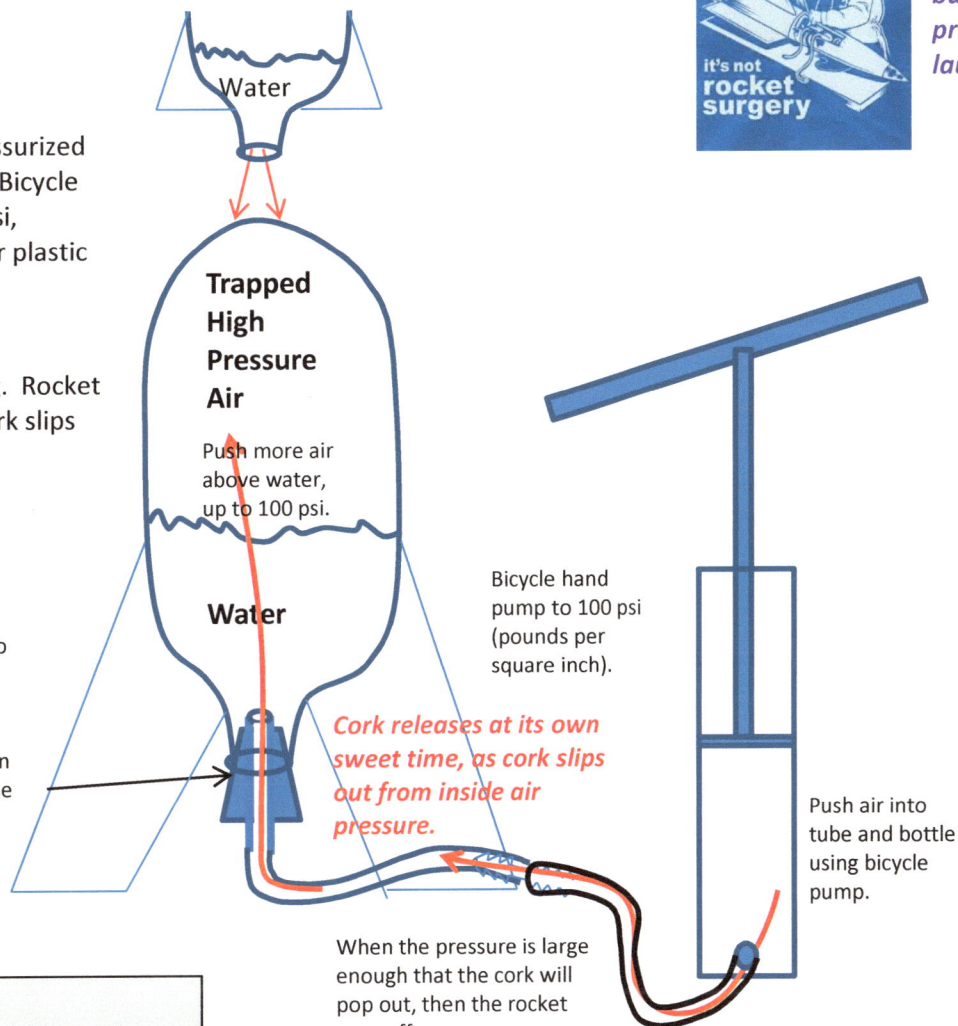

Cork releases at its own sweet time, as cork slips out from inside air pressure.

Bicycle hand pump to 100 psi (pounds per square inch).

Push air into tube and bottle using bicycle pump.

When the pressure is large enough that the cork will pop out, then the rocket goes off.
When the cork is pushed in harder, the pressure to pop out will get larger.

Pre-made Kit versions

Bung pushed into wide nozzle

Pressure pushes bung out, **<30 psi.**

Hardware store follies:
1st trip: get PVC pipes and rubber cork
...rubber cork ripped apart by drill bit.
2nd trip: buy more cork

Simple corks, without a launch pad, are the quick and dirty way to launch a water bottle, but the pressure is low.

Back Cover

Hope you had a 'blast'.

Court Rossman first saw and built water bottle rockets, as an adult, while helping with Cub Scouts. Water rockets seemed like a great introduction to hand's-on rocket building. He was impressed that there was a basic rocket to demonstrate the concept, and that the kids could build the rocket themselves. Court certainly learned himself as a volunteer from doing the activity.

This book relates the water bottle rockets and toy chemical rocket experiments to real uses and applications. Rockets have launched humans into the 21st century. The concepts are all in the main part of the book.

Court is participating in the emphasis on hands-on learning. Science kits are readily available these days, and that is great. He just wants to help that trend, so the next generation has practical knowledge and creativity.

Court has also published 'Gamut of Speedy Rockets', 'Pinewood Derby Cars and Real Cars', and 'Magnets, Motors, and Generators'. Court has a life-long interest in physics and science, and has a Ph.D. in physics, and works on rocket sensors through his day job.

Thanks to the Cub Scout Pack that staged some Water Rocket activities. Thanks to Brian Stevens who built a bump-seal PVC launder and launched the rocket activity, where scouts made their own rockets. Thanks to Elizabeth Hamparian for suggesting the appendix teacher's guide to group experiments and suggestions to simplify the book, although she would have preferred even simpler. Thanks to Stephen McDowell for taking some launch videos. Thanks to family, in particular Peach Rossman for giving always critical feedback (and some drawings) and Richard Lafleur for feedback. Thanks to Allan Rossman for drawing some sketches. Thanks to Wayne and Kim Rossman for simplifying the cover pages and general formatting. Thanks to people who posted pictures on the internet, many of which are used in this book.

Model of Dr. Goddard's first launch liquid fuel rocket

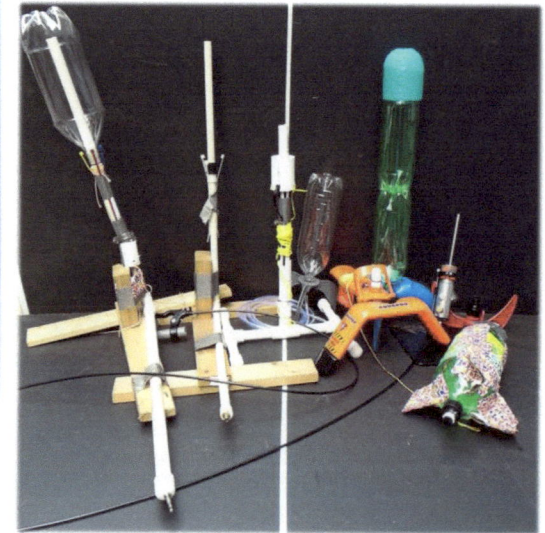

Launchers used in this book

Astronaut jokes:
https://funkidsjokes.com/astronaut-jokes/
https://upjoke.com/astronaut-jokes

References:
Youtube channel: Kurzgesagt – in a Nutshell
https://upjoke.com/astronaut-jokes

www.ingramcontent.com/pod-product-compliance
Lightning Source LLC
Chambersburg PA
CBHW041059210326

41597CB00004B/139

9780578353340